CANAD

REGIONAL ENVIRONMENTAL ISSUES MANUAL

Bringing Environmental Issues Closer to Home

TONY LEIGHTON

SAUNDERS COLLEGE PUBLISHING

Harcourt Brace College Publishers

Fort Worth Philadelphia San Diego New York
Orlando Austin San Antonio Toronto
Montreal London Sydney Tokyo

Regional Environmental Issues Manuals are unique supplements designed to accompany Saunders College Publishing environmental textbooks including *Environment*, by Raven, Berg, Johnson, and *Environmental Science* by Arms. Each *Regional Environmental Issues Manual* is designed to promote grassroots awareness of local environmental issues, problem-solving analysis, and verbal and written discussion of topics that pertain to specific environmental issues in Canada and the United States.

Regional Environmental Issues Manuals attempt to present a range of views on select environmental issues in a non-biased approach within limited space constraints. The goal of the manuals is to encourage student individual analysis of complex issues which go beyond the scope of the publication. Thought-provoking questions, commentary, and readings have been included to stimulate students to investigate the issue in further detail beyond the manual's presentation.

We invite your comments, ideas, and feedback. After reading the manual, please complete and return the Professor and Student Comments Form located in the back of the manual.

Copyright ©1993 by Harcourt Brace and Company

All rights reserved. No part of this publication may be reproduced or transmitted in any form or by any means, electronic or mechanical, including photocopy, recording, or any information storage and retrieval system, without permission in writing from the publisher.

Requests for permission to make copies of any part of the work should be mailed to: Permissions Department, Harcourt, Brace and Company, Publishers, 8th Floor, Orlando, Florida 32887.

Printed in the United States of America.

Leighton: Canadian Regional Environmental Issues Manual

ISBN 0-03-097143-8

345 256 987654321

CANADIAN REGIONAL ENVIRONMENTAL ISSUES MANUAL

Table of Contents

PROFILE OF THE CANADIAN REGION 1

1. ACID DEPOSITION

Introduction 2
Questions and Concerns for Readings 1 through 3 3
Reading 1 "The Process of Acidification" 4
Reading 2 "Ecological Effects of Acidic Deposition" 6
Reading 3 "Socioeconomic Consequences of Acidification" 10
Questions and Concerns for Reading 4 12
Reading 4 "The Little Lakes That Could" 12
Questions and Concerns for Reading 5 17
Reading 5 "The Real Story about Canada's Acid Rain Campaign in the U.S. – A Symptom of Our Malaise" 17
Questions and Concerns for Reading 6 29
Reading 6 "Signs of Progress" 30
Discussion 31

2. THE ATLANTIC COD FISHERY

Introduction 32
Questions and Concerns for Readings 1 through 4 33
Reading 1 "A Fine Kettle of Fish" 34
Reading 2 "Net Losses" 40
Reading 3 "Sky Patrols Over the Grand Banks" 46
Reading 4 "Can We Stop the Foreign Plunder of Grand Banks Fish Resources?" ... 50
Reading 5 "Haddock Getting Healthier" 54
Questions and Concerns for Reading 6 55
Reading 6 "Fishing For Trouble" 55
Questions and Concerns for Reading 7 56
Reading 7 "The Squid, the Cod and Who We Are" 57
Discussion 59

3. ARCTIC POLLUTION

Introduction 60
Questions and Concerns for Readings 1 through 4 62
Reading 1 "The Not-So-Pristine Arctic" 62
Reading 2 Excerpts from "Arctic Pollution: How Much is Too Much?" 66
Reading 3 Excerpted from "Doing the Great Slave Samba" 76
Reading 4 Excerpted from "Arctic Oil Resources and Sustainable Development" 78
Reading 5 "Nuclear Waste: A Time Bomb for the Arctic?" 82
Questions and Concerns for Reading 6 85
Reading 6 "Stormy Weather" 85
Questions and Concerns for Reading 7 88
Reading 7 "Citizens for the Environment: Is the Sword Too Blunt?" 88
Questions and Concerns for Readings 8 & 9 89
Reading 8 "Sizing Up the Strategy" 90
Reading 9 "Underwater Noise" 91
Discussion 93

4. TOXIC POLLUTION IN THE GREAT LAKES

Introduction 94
Questions and Concerns for Readings 1 & 2 95
Reading 1 "Great Lakes Clean-up at a Critical Turning Point" 96
Reading 2 "Toilet Bowl Current Stops Swimmer" 102
Questions and Concerns for Readings 3 & 4 103
Reading 3 "Lake Superior and Zero Discharge — The Effort by Government" 103

Table of Contents

Reading 4 "Lake Superior and Zero Discharge — Environmentalists Respond" 106

Questions and Concerns for Reading 5 107

Reading 5 "Sixth Biennial Report Emphasizes Virtual Elimination of Persistent Toxic Substances" 107

Reading 6 "Municipal Waste, Nutrient Enrichment and Eutrophication" 109

Questions and Concerns for Readings 7 & 8 112

Reading 7 "St. Lawrence Water Deplorable, Study Says" 112

Reading 8 "Beluga Contamination Confirmed" 113

Reading 9 "Groups Demand Freeze on Kodak's Discharge" 114

Discussion 115

5. FORESTRY

Introduction 116

Questions and Concerns for Readings 1 & 2 117

Reading 1 "Heartwood" 118

Reading 2 "Ontario's Search for a New Way in the Woods" 126

Reading 3 "The Winds of Change Blow Through Community Forests" 130

Reading 4 "Small Is Beautiful" 133

Discussion 135

6. THE JAMES BAY HYDROELECTRIC PROJECT

Introduction 136

Questions and Concerns for Reading 1 138

Reading 1 "The Geese Have Lost Their Way" 138

Questions and Concerns for Reading 2 142

Reading 2 "James Bay: Open Letters to Members of NRDC" 142

Questions and Concerns for Reading 3 146

Reading 3 "Megawatts or Negawatts" 147

Questions and Concerns for Reading 4 148

Reading 4 "Secret Pacts Cost Billions, Hydro Québec Studies Show" 148

Discussion 149

7. THE BUILT ENVIRONMENT

Introduction 150

Questions and Concerns for Reading 1 152

Reading 1 "How Green Can the Built Environment Become?" 152

Questions and Concerns for Reading 2 156

Reading 2 "R-2000" 156

Questions and Concerns for Reading 3 161

Reading 3 "Home Power" 161

Questions and Concerns for Readings 4 through 7 163

Reading 4 "The Issues: Occupant Health" 164

Reading 5 "The Issues: Water Quality" 166

Reading 6 "The Issues: Light, Sound and Radiation" 167

Reading 7 "The Issues: Embodied Energy" 168

Discussion 169

8. BIODIVERSITY

Introduction 170

Questions and Concerns for Reading 1 171

Reading 1 "The World's Food Supply at Risk" 172

Questions and Concerns for Reading 2 179

Reading 2 "Seeds of Change: Pat Roy Mooney" 179

Questions and Concerns for Reading 3 183

Reading 3 "Apple's Seeds" 183

Reading 4 "Ancient Foods for Future Diets" 188

Discussion 191

9. MINING UPDATE

Update on Raven/Berg/Johnson *Environment*, p.187 192

TABLE OF CONTENTS

1. Acid Deposition
This issue shows why acid deposition is a serious environmental problem in Canada; how Canadian scientists have determined the extent of the damage; and the role activists played in pressuring the federal government to regulate sulphur dioxide emissions in Canada and exert diplomatic pressure on the United States to do the same.

2. The Atlantic Cod Fishery
This issue examines the decline of Canada's Atlantic cod fishery. In 1992, a two-year moratorium was placed on fishing for the northern cod, one of the country's most important natural resources. This is a classic example of a "tragedy of the commons," where a common resource is shared by competing interests and inevitably ruined because no agreement can be reached on managing it sustainably.

3. Arctic Pollution
This issue examines the impact of environmental contamination on Arctic ecosystems and native peoples. The Far North has always been considered unspoiled and pristine. Recent research shows it is not only becoming polluted, but that the Arctic is especially vulnerable to pollution because of its climate and extreme latitude.

4. Toxic Pollution in the Great Lakes
This issue examines how the Great Lakes, the world's single most important freshwater resource, has become polluted with toxic chemicals that threaten the survival of many animal species and the health of people who live within the Great Lakes basin.

5. Forestry
This issue examines the forest industry in Canada. It looks at the relationships between government and forestry companies; how trees are harvested and some possible alternatives to clearcutting – the most destructive form of harvesting; and the importance of preserving biodiversity within Canada's forests.

6. The James Bay Hydroelectric Project
This issue examines the impact of the first phase of the James Bay hydroelectric project on the native people who live in the region; it shows the position of the Canadian government and Hydro Québec on the proposed second phase of the project; and why many people are opposed to the building of more dams in northern Québec.

7. The Built Environment
This issue examines the environmental efficiency of houses and other buildings, and how Canada is a world leader in the development of environmentally progressive housing.

8. Biodiversity
This issue examines the value of biodiversity among plant species, and how the continuing worldwide depletion in biodiversity threatens global food supplies and the development of valuable plant byproducts such as medicines. We also look at the work of two Canadians who are helping to defend and rebuild the Earth's dwindling plant genetic resources.

Canadian Profile

Tony Leighton

Introduction

Acid Deposition

This issue shows why acid deposition is a serious environmental problem in Canada; how Canadian scientists have determined the extent of the damage; and the role activists played in pressuring the federal government to regulate sulphur dioxide emissions in Canada and exert diplomatic pressure on the United States to do the same.

During the 1980s, "acid rain" was perhaps the single most worrisome symbol of environmental degradation for Canadians. "Few might understand fully the causes and effects of the problem," writes Dr. Andrew Forester, in *The State of Canada's Environment*, "but one fact was alarmingly clear: If something was wrong with the rain, it would affect everyone."

For this reason, acid rain is also the environmental front on which the most progress has been made – scientifically, legislatively, and through the work of well-organized activists. (Since rain is only one of the vehicles that carry acidic pollution – the others being snow, cloud, fog, dust, and gases – the correct term being used these days is "acid deposition.")

The prime acid-forming pollutants are sulphur dioxide and nitrogen oxides. In Canada, somewhere around 4.6 million tonnes of sulphur dioxide and 1.8 million tonnes of nitrogen oxides are emitted each year from metal smelters, power plants, and fossil-fuel-burning vehicles (based on 1980 figures). In the United States the equivalent amounts are 24 million and 20 million tonnes. Studies have shown that approximately 50% of the acid deposition that falls in Canada originates in the United States, transported from heavy industries in the U.S. Midwest to Canada on prevailing air currents. This makes the rain in eastern Canada 10 times more acidic than it would be without the pollution.

Acid deposition causes a number of unfortunate biological and physical changes. Most of us are aware that increased acid in lakes and streams destroys fish populations. It also draws heavy metals from soil and plumbing fixtures, releasing them into the ecosystem, and it corrodes brick and stone. It may also be causing a reduction in the nutrient value of agricultural soils, forest "dieback," and an impairment of human lung capacity.

Canada is susceptible to acid deposition because 46% of its land mass (primarily land east of Manitoba) is unable to counteract acid pollution with natural alkalines. The vast area of northern Ontario and Quebec known as the Canadian Shield, which is characterized by thin soils and protruding granite bedrock, has an especially low tolerance for acid deposition. It also happens to contain a large portion of Canada's lakes and wetlands.

The amount of acidic sulphur falling on eastern Canada has declined steadily since the 1970s, due mainly to improved industrial smokestack-scrubbing technologies required by government regulations. In 1985, federal/provincial negotiations through the Canadian Council of Resource and Environment Ministers yielded an agreement on emission-reduction targets to each of the seven provinces from Manitoba eastward. Allowable emissions of sulphate were roughly to be halved between 1980 and 1994. In the United States, important amendments to the *Clean Air Act* were passed in November of 1990, regulating the control of sulphur dioxide and nitrogen oxide emissions. By 1995, total annual sulphur dioxide emissions will be reduced in the U.S. by 4.5 million tonnes; by 2000, they will come down an additional 4.5 tonnes. When these legally-mandated reductions are made, and if proposed reductions are achieved in Canada, sulphate deposition should drop below the 20 kilogram-per-hectare-per-year targets set by Environment Canada.

But it's premature to celebrate. While the Canadian provinces are working toward emission-control objectives, they are not bound by legislation like the *Clean Air Act*. It remains to be seen whether the large corporations seemingly most responsible for the pollution, such as mining firms, steelmakers and electric-power utilities, will invest in the necessary technology to cut their emissions during the recession. By 1993, they had only reached 40% of the overall nine-year target for 1994.

Studies show that even if full reductions are made in both countries, eastern Canada will still suffer from the acid in its skies. The target "loading value" of 20 kg/ha per year represents an acidity level that can be withstood by moderately sensitive water bodies. However, more biologically sensitive habitats might only be restored at 8-12 kg/ha. A Royal Society report issued in 1992 states that a 50% reduction in emissions will likely result in a 50% recovery in many of the affected lakes and streams. We would only be halfway back.

In the end, the acid deposition problem is not, at its core, about the affordability of smokestack scrubbers or how many tonnes of sulphur dioxide a mining company should be allowed to spew into our common atmosphere. It's a matter of the efficiency with which we use energy and the energy sources we choose in the future.

The least expensive way to prevent the sulphur dioxide and nitrogen oxide emissions that cause acidic deposition is to burn less fossil fuel by making our industries, homes, and automobiles more energy efficient. As leading American electric utilities have found, investing in conservation through the purchase of super-efficient lightbulbs, or refrigerators, or structural insulation is less expensive than trying to put filters on old

Tony Leighton

fossil-fuel plants or build new plants. The net effect of energy efficiency is lower costs for the utility (and its customers), reduced sulphur dioxide emissions into the environment, and reduced acidic deposition.

The next logical strategy is to invest in the development of alternative sources of energy – renewable sources that do not produce sulphur dioxide and nitrogen oxides. Unfortunately, governments in North America have not yet taken this obvious step. While the *Clean Air Act* in the United States is encouraging, as is President Clinton's new program to encourage the use of natural gas, other government initiatives fall far short. The U.S. National Energy Strategy, proposed by the Bush administration before leaving office, relies heavily on fossil fuels while offering only nominal support to alternative energy sources like solar, wind, geothermal, biomass, and hydrogen – all increasingly viable these days.

The point is elementary: If we are to have clean skies, free of acidic pollution, we must first use less energy, then cleaner energy.

Introduction

Questions and Concerns

READINGS 1 THROUGH 3

1. Ask yourself why "acid rain" became the most important environmental issue in the minds of Canadians during the 1980s.

2. Acidic deposition, carried from the United States to Canada, is a classic example of pollution knowing no borders. Think of other examples elsewhere in the world where one country's pollution is affecting another country's environment.

3. It has been suggested by a number of experts that forest "diebacks" in parts of Quebec, Ontario, and the U.S. Northeast can be attributed to acid deposition. As you read about the effects of acidic deposition on soil and trees, ask yourself why this is likely.

4. Many people believe that only lakes are affected by acidic deposition. Make note of the other deleterious effects it has on water, land, animals, plants, buildings, and human beings. Take special note of one author's interesting contention that "symbolic values" are at risk; that acid deposition has a demoralizing effect on people. What reasons does he give?

THE PROCESS OF ACIDIFICATION

Recognizing the problem

More than a century ago, Robert Angus Smith, a British chemist, coined the term "acid rain" (Smith 1872), and, despite the limitations of a technology that was primitive by today's standards, he demonstrated that smoke and fumes contained substances that caused significant changes in the chemical composition of precipitation, changes that could be detected not only close to the source, but also "in the fields at a distance." He also identified some of the harmful effects of "acid rain," such as the bleaching of coloured fabrics, the corrosion of metal surfaces, the deterioration of building materials, and the dieback of vegetation.

In Canada, acidic deposition was not identified as an environmental concern until the 1950s. Working at Dalhousie University in 1955, Dr. Eville Gorham attributed the "abnormal acidity" that he detected in precipitation, and in the water of Nova Scotia lakes, to airborne pollutants which, he theorized, might come from distant sources. Surface water surveys and monitoring that had been undertaken by the Canada Department of Mines and Technical Surveys (Thomas 1953) provided a valuable baseline for early documentation of the trends and consequences of acidification.

In 1966, Dr. Harold Harvey, at the University of Toronto, detected severe losses among fish populations in the lakes of the La Cloche Mountains, southwest of Sudbury, Ontario, and ascribed these losses to acidification of the waters by acid rain (Beamish and Harvey 1972). Two large research programs, the International Joint Commission's reference reports on the upper Great Lakes (International Joint Commission 1982) and the Sudbury Environmental Study undertaken by the Ontario ministries of Environment and Natural Resources (Ontario Ministry of the Environment 1979), both produced reports that identified this state of abnormal chemistry in precipitation falling on Ontario lands and waters.

Such research findings, backed by others from abroad, prompted the convening of a symposium in 1975 to address the "Atmospheric Contribution to the Chemistry of Lake Waters" (Matheson and Elder 1976). In May 1976, Environment Canada directed the formation of a committee of scientists to study the long-range transport of airborne pollutants (LRTAP). The report of this group (Whelpdale 1976) stated that:

> "...in Canada, the potential for problems of this nature exists. We have the meteorological conditions which are conducive to transport from source regions in Canada and the United States to regions of the country which have sensitive soils, waters, fish and forests."

What is "acid rain"?

In the popular media, the term "acid rain" has been used very loosely. Technically, acid-forming and acidic pollutants are transported and deposited via the atmosphere, not just in the form of rain, but as snow, cloud, and fog ("wet" types of deposition) as well as gases and dust ("dry" types) during dry periods. When we want to refer broadly to all these different forms, the term "acidic deposition" is more appropriate than "acid rain."

Around the world, normal precipitation is slightly acidic, with a pH of approximately 5.6–5.0. This modest amount of acidity comes from the absorption of carbon dioxide and other acidic material of natural origin. In some areas, the pH value of precipitation may exceed 5.6, as in the Prairie provinces, where wind-blown alkaline soil materials actually reduce the natural acidity levels.

However, precipitation over eastern North America, and much of Europe as well, has been found to be 10 times more acidic than normal, largely because of the incorporation of airborne sulphur dioxide (SO_2) and nitrogen oxides (NO_x). These human-caused gases react in complex ways to form sulphuric and nitric acids in the atmosphere — in rain, snow, and cloud droplets. Oxides of sulphur and nitrogen may reach the earth directly as gases or dry particulates and react to create acidic conditions. Thus, both pollutants add acidity to the clouds and precipitation; cloud droplets have had pH levels as low as 2.6 when measured at Whiteface Mountain in New York state (Castillo 1979).

Sources of acid-forming pollutants

Although natural sources of sulphur oxides and nitrogen oxides do exist, more than 90% of the sulphur and 95% of the nitrogen emissions occurring in eastern North America are of human-made origin. The largest sources of these pollutants are the smelting or refining of sulphur-bearing metal ores and the burning of fossil fuels for energy.

In 1985, about 70% of the sulphur dioxide emissions in the United States came from coal- or oil-fired electrical generating stations; in Canada, about 50% was emitted from ore smelters and about 20% from generating stations burning fossil fuels. Sulphur emissions in both countries tend to be concentrated in a relatively few locations. By contrast, the sources of nitrogen emissions are widely distributed. About 40% of nitrogen oxides come from transportation (cars, trucks, buses, trains), about 25% from thermoelectric generating stations, and the balance from other industrial, commercial, and residential combustion processes.

In 1980, which has served as the reference year for tracking emissions, sulphur dioxide emissions were estimated to be

Reproduced from: Government of Canada, 1991, *The State of Canada's Environment*.

4.6 million tonnes in Canada, and 24 million tonnes in the United States. Emissions of nitrogen oxides in the two countries were 1.8 million and 20 million tonnes, respectively.

Sulphate deposition

Airborne acidic pollutants are often transported thousands of kilometres from their point of origin before being deposited. Because the movements of air masses and storms tend to follow systematic patterns, pollutants from particular locations are, in effect, channeled away and generally fall on the same recipient areas. In eastern North America, weather patterns generally travel from southwest to northeast. Thus, pollutants emitted from sources in the industrial heartland of the midwestern states and central Canada regularly fall on the more rural and comparatively pristine areas of the northeastern United States and southeastern Canada.

Because sulphur dioxide emissions are monitored at their source, they can be measured for both Canada and the United States on a provincial or state basis. Computer-based simulation models use emission data to approximate where the pollutants will be transported. These hypotheses are then confirmed by comparison with the actual amount of sulphate deposition recorded at weather monitoring stations across Canada and the United States. Modelling enables researchers to estimate how much of the sulphate measured at a given recording station originates in a particular region. By this method it has been estimated that about 50% of the sulphate deposited in Canada is derived from sources in the United States (RMCC 1990, part 3).

By correlating the amount of acidic deposition in a given region with observed damages to ecosystems, it becomes possible to describe the responses that can be expected from differing levels of acidic emissions and subsequent deposition.

A continental network of weather stations has been monitoring acidic deposition since about 1979, long enough to indicate changes or trends that have occurred over the past decade. Sulphate deposition in precipitation, though still at undesirable levels in many parts of eastern Canada, has declined at an approximately steady rate. When averaged over eastern North America, the decline in sulphate deposition is seen to be in nearly direct proportion to the decline in emissions. This observation supports the proposition that levels of acidic pollutants in the atmosphere can be reduced by reducing emissions.

ECOLOGICAL EFFECTS OF ACIDIC DEPOSITION

Once it enters an ecosystem, acidic pollution influences a wide variety of ecological processes. The water cycle in nature is analogous to the circulation of blood in a living creature. When blood chemistry is altered beyond acceptable limits, the basic life functions of the organism are impaired. Acidification throughout the water cycle can have a similar effect on entire ecosystems, and therein lies its importance as a major threat to the health of the environment on both regional and global scales.

Compared to the slow pace of natural geophysical and evolutionary processes, the systemic changes induced by these airborne contaminants have been virtually instantaneous. Such abrupt and radical events have the potential to wreak havoc on ecosystems whose rhythm of evolution and adjustment is often measured, not in years and decades, but in centuries and millennia.

Sensitivity of ecosystems

The effects of abnormally acidic water on rocks, soils, plants, and animals combine to determine the ultimate response of ecosystems, and especially the fresh water that they contain, to acidic deposition. Rock weathering may be accelerated, soil nutrients and trace metals may be altered, plant growth may be impaired, and the ability of habitats to sustain living organisms may be diminished. The degree to which a particular area is unable to resist the damaging effects of acidic deposition is termed its sensitivity to acidification.

The sensitivity of a region or ecosystem is directly related to its capacity to neutralize or counter excess acidity. If, following an episode of acidic deposition, adverse effects are minimal and conditions return quickly to near normal, the ecosystem is not considered to be sensitive. If, on the other hand, the ecosystem suffers severe damage because it has little ability to neutralize acidic pollutants, it is deemed to be highly sensitive.

The capacity of any region to neutralize acidity is determined primarily by the characteristics of its soils and bedrock (Cowell and Lucas 1986). A model based on available information from terrain, soil, and ecological land inventories ranks Canadian landscapes in three sensitivity classes — high, moderate, and low — with regard to their potential to withstand acidic deposition. A large proportion of the landscapes with low capacity for reducing acidity, and thus with high sensitivity, lie in the Canadian Shield. This area is dominated by granitic bedrock and thin, poorly developed soils, and it contains a large proportion of Canada's wealth of lakes and wetlands. In all, the highly sensitive category encompasses about 4 million square kilometres, about 46% of the country's surface area.

Clearly, the occurrence of high levels of acidic deposition in regions with low neutralizing capacity is most likely to seriously damage ecosystems. By comparing the areas with high levels of wet sulphate (sulphate measured in precipitation) deposition and the are as of high sensitivity, it is easy to see where the two conditions coincide. It is there that there is the greatest cause for concern. In 1980, more than 10 kg/ha of sulphate was deposited over sensitive landscapes in Ontario, Quebec, the Atlantic provinces, and, to a lesser extent, southwestern British Columbia. Significant, and in some cases severe, ecological responses to acidic deposition, in the form of damage to lakes, are found in these regions.

Terrestrial responses

In wooded areas, the foliage of the upper canopy will intercept and collect some precipitation, including its acidic burden. The leaves also filter out dry particulates and gases and slow the rate at which acidic pollutants reach the ground. When such substances do enter the soil, whether in free-fall from the air or by trickling down tree trunks and plant stems, they elevate the acidity of the groundwater and alter natural soil cycles, generally diminishing the availability of soil nutrients and accelerating the process of chemically wearing away the soil. Heightened acidity also induces new reactions to which the ecosystem is not adapted. Principal among these is an increased demand for, and release of, the so-called "basic" (i.e., alkaline) elements, such as calcium and magnesium, into the water in the soil.

Initially, this leaching of basic elements away from plant roots and soil particles serves to buffer and neutralize the excess acidity. However, many Canadian soils have minimal reserves of basic elements. As the rate of leaching exceeds the availability of these buffering materials, the water remains acidic. Metals that are toxic at elevated concentrations, such as aluminum, begin to dissolve and are transported by the groundwater flow to streams and lakes. In some instances, the soils may undergo such radical modifications that it can no longer sustain certain established plant populations.

It is thought that acidic and other atmospheric pollutants may damage vegetation not only through modification of nutrients in the soil, but also through direct contact with plant tissues. Controlled experiments have shown that these substances can alter the waxy, protective surfaces of leaves and the capacity of guard cells to function. The implications of these changes are not yet well understood, but they

Reproduced from: Government of Canada, 1991, *The State of Canada's Environment*.

are believed to decrease disease resistance and may also inhibit germination and reproduction of the species affected.

Plants may also be affected by indirect modifications of the ecosystem. High soil concentrations of aluminum can prevent the uptake and use of nutrients by plants. The leaching of soil nutrients and the release of toxic aluminum, as a result of acidification, may result in nutrient deficiencies or imbalances in plants. The damage is not limited to individual species of plants and animals, but in many instances extends to the biological communities in which they exist. Increased availability of aluminum in soils has been implicated as a cause of forest declines in Europe and North America (RMCC 1990, part 5).

Forests

Extensive forest declines and diebacks have occurred in Europe (Krause *et al.* 1986) and on the higher slopes of eastern mountains in the United States (Johnson *et al.* 1986). Much of the evidence points to acidic deposition as a cause of this damage.[1] In eastern Canada, 96% of the land with high capability for forestry is subject to acidic deposition in excess of 20 kg/ha per year (Lynch-Stewart *et al.* 1987). Approximately 15 million hectares of hardwood and mixed-wood forests are exposed to significant levels of sulphate and nitrate deposition (RMCC 1990, part 5). In recent years, important instances of dieback and declines in growth rate have been noted in sugar maple groves in parts of Canada that receive high levels of these and other air pollutants, such as ozone.

[1] In addition to acidic pollutants, Canada's forests endure the combined pressures imposed by other stresses: climatic extremes and changes, invasions by insects and diseases, and the harvesting practices of the forest industry. All of these stresses modify forest health and productivity. In view of this complex situation, it has not been possible to establish the exact role of acidification in forest decline, nor to develop critical deposition levels at which damages are believed to become important (RMCC 1990, part 5). However, the geographical coincidence of forest decline in regions of elevated acidic deposition strongly suggests that such links exist. Nitrogen oxides also contribute to the formation of ozone in the air near the Earth's surface; at low elevations, ozone may act in conjunction with acidic deposition to harm vegetation, including forest growth.

In Ontario, significant symptoms of decline are occurring throughout the range of hardwood forest, especially in northern locations. The declines, most notable over the past 30 years, coincide with a period of rapidly increasing industrialization and urbanization across much of the province. A survey of more than 2 million hectares of sugar maple stands in Quebec, conducted between 1985 and 1987, revealed the following damages: 3% of the area exhibited severe damage (>25% defoliation); 47% exhibited moderate damage (11–25% defoliation); 50% showed only marginal symptoms (<10% defoliation). One disquieting aspect is that indications of decline among maples increased noticeably during the two years of the study.

White birch stands in eastern Canada may be in even more serious condition. A survey in the Bay of Fundy region in 1988 revealed that nearly all specimens were affected to some extent, and about 10% of the trees were already dead (RMCC 1990, part 5). Similar patterns of decline have been observed in stands of white birch on the shores of Lake Superior, another area where acid fog occurs frequently. The contribution of acidic pollutants to forest health problems is a subject of ongoing study by federal and provincial forestry departments.

Farmlands

Acidic deposition alters soil chemistry on farmlands as well as forests. The application of agricultural lime as a neutralizer has become a standard practice in many parts of eastern Canada. The precise effect on crop yields of exposure to ambient levels of acidic pollutants is unknown. Experimental work in this field is ongoing but, partly because of complex interactions with other pollutants such as ozone, it has not yet been possible to establish critical levels of acidic deposition in

[2] Within that total, 54% of the lakes covered less than 1 ha. Although such tiny lakes contain a relatively small proportion of the total supply of surface fresh water, they provide essential habitat for waterfowl and aquatic organisms (McNicol *et al.* 1987). Any influence that may disturb the balance of these vital ecosystems must be viewed with concern.

relation to crop damage.

Terrestrial wildlife

The effects of acid rain on terrestrial wildlife are even harder to assess. Generally, they would occur indirectly, as a result of pollution-induced alteration of habitat or food resources. Loss of forest cover, for example, could reduce available habitat for many species of birds. European studies have indicated that alteration of the cycle of calcium nutrients in soils and waters causes dietary deficiencies and may result in eggshell thinning and poor reproductive success. In Canada, evidence suggests that lichens and other plants take up and concentrate toxic metals, which may in turn accumulate in the livers of large ungulates such as caribou and moose. The amount of cadmium in moose livers has been found to be high enough that humans should avoid or limit consumption of this organ (RMCC 1990, part 5).

Aquatic responses

Canada has a wealth of fresh water. Lakes, streams, and wetlands cover about 7.6% of the country's surface area, and the volume of groundwater is many times greater than the volume of surface water. Recently, a satellite study of water bodies (Hélie and Wickware 1990) was conducted for the region of Canada that is most affected by acidic pollution, that is, the area east of the Manitoba–Ontario border and south of 52° north latitude. In this region more than 775 000 water bodies larger than 0.18 ha were identified.[2]

The task of gathering and evaluating data on all these lakes, ponds, wetlands, streams, and rivers is immense: at present, chemical analyses are available for only about 1% of the total. Even though some of the data were collected as early as the 1950s, systematic surveys have been conducted only during the last decade. Fortunately, this limited sampling base offers a reasonably good representation of the chemistry of the entire water resource, although there is a possibility that acidic and highly sensitive waters may be underrepresented.

Water chemistry is determined by the interaction of precipitation with the geological and biological conditions that exist over a given region. Surface waters that have been derived from unpolluted precipitation, and that flow through weathered soils and rocks before reaching streams and lakes, will contain dissolved basic elements, like calcium and magnesium, along with a chemical equilibrium of bicarbonate (HCO_3).[3] The concentration of these chemicals is determined largely by the geological properties of the watershed. Waters low in dissolved minerals are termed "soft," whereas those with high concentrations are termed "hard."

Most of the unpolluted waters in eastern Canada are naturally very soft (LRTAP Working Group 1989) due to the bedrock and soil characteristics of the Canadian Shield. The same characteristics that produce soft waters also define areas that are sensitive to acidic deposition. Such waters, unless otherwise polluted, will have a near-neutral acidity, ranging between pH 6 and somewhat greater than pH 7, and will contain only trace amounts of sulphate or nitrate (RMCC 1990, part 4).

When pollutants containing sulphuric and nitric acids are introduced into soft waters, the acids are first neutralized or buffered by the bicarbonate and, to some extent, by the release of other elements, including aluminum. The evidence of such early acidification is most easily detected by the loss of bicarbonate and the appearance of sulphate in the water. At this stage, any nitrate that is deposited is usually assimilated as a nutrient by plants. It contributes directly to acidification of the water only when vegetation and other aquatic organisms cannot assimilate all that is deposited. When all the available bicarbonate in the water body has been exhausted, the waters become significantly acidic, with a pH of 5.0 or lower. Researchers can measure the degree to which acidification has taken place in a lake by determining the relative amounts of bicarbonate and sulphate in samples of its water.

[3] For a more detailed discussion of water chemistry, readers should consult one of the standard texts, such as Stumm and Morgan (1970).

To determine how widespread the acidification of waters in southeastern Canada might be, more than 8 500 lakes have been surveyed and analyzed for their chemical content (RMCC 1990, part 4). In areas where the ratio of bicarbonate to sulphate is less than one part bicarbonate to one part sulphate, the waters are experiencing serious acidification. If the ratio has fallen to 1:5 (expressed as 0.2) or lower, it would be expected that many acidic water bodies with a pH below 5 would be found. As the map indicates, large portions of Ontario, Quebec, and the Atlantic provinces fall into these classes.

In addition to the long-term process of acidification, short-term events can have severe consequences for small bodies of water. In much of Canada, a large proportion of the total annual precipitation is deposited as snowfall, or in severe rainstorms. Sudden spring snowmelts and heavy storm runoffs have the potential to introduce large quantities of acidic pollutants into the surface water system in a very short time, introducing massive pulses of increased acidity to local streams and lakes. On occasion, these temporary pulses or "acid shocks" have killed large numbers of fish (Marmorek *et al.* 1987) before the sudden input of acidity could be buffered. Because acidity in this form does not have time to be assimilated gradually into the environment, it is of particular concern, even in areas where long-term acidification has not yet reached levels that would normally be considered serious.

Aquatic organisms

Changes in the flora and fauna of aquatic ecosystems are often the first and most immediate indicators of acidification in the lakes of eastern Canada. The interactions between living organisms and the chemistry of their water habitats are extremely complex. If a species or group of species increases or declines in number in response to acidification, then the ecosystem of the entire water body is likely to be affected. Reactions of organisms to stresses such as acidification have been termed "dose-response reactions" (i.e., a certain dose of an acidifying pollutant induces a certain reaction).

By causing direct changes in a single component of the food web of an ecosystem, acidification can indirectly modify predator–prey relationships throughout the entire system. These indirect effects make accurate assessment of biological damage difficult, as it is necessary to understand which effects are caused directly by acidic pollution to draw dependable conclusions. By the time that a major biological response, such as the loss of a given fish stock, has become evident, serious damage may already have occurred in the aquatic ecosystem.

Knowledge relating biological responses to habitat acidification has been obtained from small-scale experiments in laboratories, from whole ecosystem experiments in which entire lakes are artificially acidified, and from extensive field surveys and monitoring. By combining these sources of information, quite a precise understanding of the effects of acidification on aquatic organisms has been obtained (RMCC 1990, part 4).

At first, the effects may be almost imperceptible, but as acidity increases, more and more species of plants and animals decline or disappear. Crustaceans, insects, and some algal and zooplankton species begin to disappear as the pH approaches 6.0, but most fish are not, as yet, affected directly. As acidity increases from pH 6.0 to pH 5.0, major changes in the makeup of the plankton community occur, and progressive losses of fish populations are likely, with the more highly valued species being generally the least tolerant of acidity. Fish declines often begin with reproductive failure, leaving a diminishing stock of aging individuals; when these fish die, their disappearance gives the impression of a sudden disaster, although the sequence of events leading to it may in fact have begun long before. At this stage of pH decline, less desirable species of mosses and plankton may begin to invade. When the pH falls below 5.0, the water is largely devoid of fish, the bottom is covered with undecayed material, and the nearshore areas may be dominated by mosses. The lake, once rich in species, has gradually become an impoverished and unhealthy ecosystem.

Although most amphibians, such as frogs and salamanders, spend the greater part of their life cycle on land, they depend heavily on temporary ponds for breeding. These small, ephemeral water bodies are highly vulnerable to the "acid shock" events associated with storms and snowmelt. In several studies, reproduction of amphibians has been shown to be seriously restricted when the acidity of their breeding habitat increases to a pH value of less than 5.0 (RMCC 1990, part 4).

To date, water birds have not been shown to be directly affected by acidification. However, if acidification of their habitat should diminish the supply of fish, invertebrates, or plants on which they feed or induce elevated levels of toxic metals in the food supply, they may suffer the consequences. Invertebrates that normally supply calcium to egg-laying females and to their growing chicks are among the first species to disappear when lakes acidify. As these food sources are reduced or eliminated, the quality of habitat declines and the reproductive success of the birds is affected. Field studies have demonstrated that the Common Loon is able to raise fewer chicks, or none at all, on acidic lakes where fish populations have been reduced (Alvo et al. 1988; Wayland and McNicol 1990). On the other hand, there are some isolated cases in which food supplies for birds increase when competing species are removed. The Common Goldeneye, a species of duck that nests by lakes and rivers in forested terrain, can better exploit insects as food when competition from fish is eliminated. Such variations make the collective influences of acidification on bird populations difficult to quantify. Generally, however, acidification remains a continuing threat to species of waterfowl that rely on a healthy aquatic ecosystem for breeding and for rearing young, in thousands of lakes across southeastern Canada (RMCC 1990, part 4).

Human health responses

Acids in rain or snow have not been observed in concentrations that would pose a direct risk to humans. However, people do inhale airborne acidic pollutants in particulate or gaseous forms. There are indications that this exposure to oxides of sulphur and nitrogen may lead to irrigation of the respiratory tract and subsequently to impaired lung function and aggravation of respiratory ailments such as bronchitis and asthma.

Humans may also be affected indirectly through exposure to increased levels of toxic metals in drinking water and food. As noted previously, increased levels of toxic metals result from the deposition of acidic pollutants into water sources, increased leaching of metals from soils and lake sediments, and increased corrosion of water pipes.

Evidence for linkages between acidic pollution and human health has been derived from epidemiologic studies of exposed human populations, from human volunteer studies, and from animal experiments (RMCC 1990, part 6). For instance, when air pollution levels in southwestern Ontario were compared with hospital admissions in the same area, positive associations were found between respiratory illness and elevated levels of sulphate, ozone, and temperature (Bates and Sizto 1989). Girls attending a summer camp in southern Ontario showed signs of diminished lung function that were correlated with episodes of acidic air pollution.

In another study, the lung function of children aged 7–12 was tested in Tillsonburg, Ontario, an area of heavy acidic pollution, and Portage la Prairie, Manitoba, where airborne acidic pollution is not severe. Lung capacity was diminished by about 2% in the more heavily polluted community (Stern et al. 1989). In a subsequent cross-sectional study of 10 towns in Canada, a similar reduction of about 2% in lung function was observed in populations of towns with higher air pollution levels. The difference appears to be due to acidic sulphate, although similar reactions have been known to occur in response to elevated levels of ozone.

Like other animals, humans can be affected by the entry of pollutants at other points in the food chain. For example, elevated levels of mercury, toxic to humans when consumed, have been found in fish taken from acidified waters. The mercury may be deposited by direct atmospheric transport or may have been released from local environmental sources by the acidification of the waters (Meger 1986).

Aluminum that is leached from soils by acidification and thereafter appears in increased concentrations in water and foods has been implicated by some researchers as having an association with Alzheimer's disease (Jackson and Huang 1983). However, other recent studies reported by Foncin in 1987 have found "...no environmental influence whatsoever on the incidence of the disease [Alzheimer's]." Thus, although the association is cause for continued concern, more research will be required to establish whether it is an indirect effect of acidification.

Although it has been well established that an increased concentration of free aluminum is associated with acidification, elevated levels of other metals have also been observed in drinking water, apparently resulting from the corrosive effect of acidified water on plumbing systems (Meranger and Kahn 1983). Levels of cadmium and lead, both of which are highly toxic, may be higher than acceptable if water from acidified sources stands in pipes for several hours. Allowing the tap to run to flush the plumbing before using the water for drinking or cooking is an effective way of avoiding health risks in such cases. In urban water systems, the problem of elevated metal concentrations is usually addressed by adjusting the pH of water at the municipal water treatment facility. In some rural or cottage settings, where water may be drawn directly from acidified sources, correcting the problem depends on the individual's own efforts. It may be necessary to install adequate filtering devices or, in cases of contamination from the plumbing, to flush the water supply system, to remove built-up contaminants.

SOCIOECONOMIC CONSEQUENCES OF ACIDIFICATION

Of all of the ecological effects of acidic deposition considered in the preceding section, the best understood relationships are with surface waters. Existing knowledge permits scientists to predict the consequences of specific levels of pollution on surface water biota, and this has been done for fish populations. To date, most researchers have calculated social and economic costs of acidic deposition using these predictions of damage to fish populations. The effects of acidification in other areas, such as agriculture, forestry, and human health, are less direct, although the evidence of a link is strong. Studies are continuing that will lead to more precise measurements of socioeconomic costs due to the effects of acidic deposition on these resources. The social and economic costs are undoubtedly great in these areas as well and provide additional justification for corrective measures, such as emission controls, even in the absence of comprehensive cause-and-effect linkages.

Although damage to natural ecosystems (and associated resource industries) is the chief threat of "acid rain," other values, including symbolic values, are also at risk. The damage that acidic deposition inflicts on maple forests is also an assault on the maple leaf, the national emblem of Canada. The disappearance of the Common Loon from acidified lakes represents an impoverishment of the Canadian wilderness experience, even for urban dwellers who may never have heard its eerie call. The corrosion of heritage buildings, structures, and monuments represents an erosion of cultural identity that may surpass in importance the monetary cost of repair or replacement. And the toll that acidic pollution exacts on human beings is not restricted to a decline in physical well-being, but includes the cost to community and individual morale of knowing that air, food, and water everywhere are affected and that, even in the wilderness, some food may be hazardous. The damage to the well-being of people who live on remote wilderness lakes and depend heavily on healthy fish populations for subsistence and for income is particularly grave.

Effects of acidic deposition on buildings, monuments, and materials

All materials used in artificial structures are subject to deterioration from normal weathering. However, deterioration has accelerated drastically since the advent of industrial pollution (Amorosa and Fassina 1983). Damage is now significant at many Canadian heritage sites, among them the federal Parliament Buildings, where even the bronze sculpture of "Canada" shows signs of serious erosion and pitting (Weaver 1985, 1987).

The National Research Council (Gibbons 1970) provided early evidence that "corrosivity" levels were highly correlated with levels of sulphur-based pollutants. Corrosion is significantly influenced by the direction or aspect of exposure relative to wind and weather, and by other factors such as frequency and duration of wetting. However, sulphur dioxides have specifically been linked with the rates of deterioration of building materials (Kucera 1983). Research efforts to establish quantitative measurements of such relationships are nearing the reporting stage in both the United States and Britain (National Acid Precipitation Assessment Program 1990; Building Research Establishment 1990).

Modern building materials have been developed to retard such deterioration. However, a great many of the historic buildings and statues that make up our cultural heritage are built of stone, brick, and masonry. The reactions of sulphur-based solutions on the surfaces and in the pore structures of these materials produce faster rates of erosion, chipping, fracture, and discolouration. Although a general lack of long-term measurements has hindered development of precise relationships, it is now well known that acidic pollutants are the cause of a rapid increase in the decay of historic structures.

Extensive repair and restoration work has been required to reverse the advanced deterioration of many of Canada's heritage buildings. These include the Art Gallery of Nova Scotia (formerly the Federal Building), the Provincial Court in Quebec City, the Bell Telephone Exchange in Montreal, St. James' Cathedral in Toronto, St. Paul's Church in Hamilton, and, to a lesser extent, the Bank of Montreal in Winnipeg (M. Weaver, conservation consultant, personal communication).

Economic costs of acidification

Traditionally, economic losses are expressed in dollar costs and are associated with cases where such values can be assigned. Although costs can be attached to some of the effects that have been discussed thus far, many of them defy this sort of interpretation. What is the real value of fish in a lake so remote that they may never encounter an angler's lure? How do we put a price on the pleasure of listening to birdsong or wandering through a healthy forest?

One of the most evident and measurable losses due to acidification is the damage to the recreational fishery. Although the methodology of evaluation remains imprecise, in 1985 an estimated 3 million resident Canadians and 680 000 nonresidents engaged in more

Reproduced from: Government of Canada, 1991, *The State of Canada's Environment.*

than 57 million angler-days of recreational fishing activity in eastern Canada. On the basis of present emission control actions, analysts have forecast that an annual increase of 5.2 million angler-days will result from improved fishing quality during the period 1986–2015. If each angler-day is valued at $73 in direct expenditures and related investment, the direct annual increase in angler spending that is projected as a result of emission/deposition controls would be about $380 million when averaged over the 30-year period (DPA Group Inc. 1987).

To take another example, an accurate estimate of the cost of acidification to the Canadian forest industry has yet to be determined. Effects that have been documented are believed to result from multiple stressors. In the absence of precisely understood relationships, a hypothetical economic value for loss of forest growth has been calculated, assuming an average reduction in growth of 5% for forest regions that are subjected to acidic deposition (Crocker and Forster 1986). Such a loss in growth, if continued on a yearly basis and converted into allowable timber harvest, would represent a potential annual economic loss of about $197 million. Nontimber losses, such as the loss of maple sugar production, the disruption of forest-based recreation, or the loss of wildlife habitat, have not been estimated. In 1987, the Canadian forest industry contributed $20 billion to the Canadian economy. In view of this, even a decline of 1 or 2% in yield on an annual basis would represent a serious threat to industry efforts to increase production (RMCC 1990, part 5).

As indicated previously, damage to heritage structures must to a large degree be assessed in terms of cultural values. We cannot easily assign a dollar value to an ancestral grave marker, to native petroglyphs on some unexplored cliff face, or to the carved stone facade of the Parliament Buildings. However, the additional annual maintenance costs for the exterior of buildings (e.g., painting) across Canada as a result of damage from acidic deposition has been estimated at about $830 million (Weaver 1985). This figure does not include damages to underlying structures, which may well be greater, albeit more difficult to assess.

The health care costs resulting from exposure to acidic pollutants cannot be separated from the overall effects of air pollution. It has been found that impairments of lung function are greater in relation to sulphate exposures than to other measures of air pollution. The 2% impairment in lung function that was observed in school children under increased levels of pollution exposure may accumulate over a lifetime, producing an age-related decline of up to 20% (Berkey et al. 1986). Hospital admissions for respiratory diseases have been linked to increased exposure to air pollutants (Bates and Sizto 1989). Although the overall economic impact of these health effects has not been determined, the combined effect of direct medical expenses and indirect losses of work time and productivity must be considerable.

Reading 4

Questions and Concerns

READING 4

1. Think about the "elegant simplicity" of the Experimental Lakes Area. Real-life experiments are being conducted on real lakes to demonstrate the real effects of pollution. Why has this been so effective?

The LITTLE LAKES that COULD

FORTY-SIX OF THEM IN NORTHWESTERN ONTARIO HAVE GIVEN CANADA A WORLD-CLASS WATER SCIENCE PROGRAM

By Barbara Robson

FEW CANADIANS WOULD HAVE HEARD OF JOHN RUDD, or his timely research, if he'd been allowed to speak up. But the federal government refused him permission to attend a New York hearing on the James Bay II hydroelectric project. Rudd, a scientist who is also a public servant, had been invited by environmentalists to talk about his work at a remote lake in northwestern Ontario. Now, thanks to the muzzling, many more North Americans on both sides of the border are curious about his startling theory.

Rudd's preliminary findings run contrary to the notion that hydroelectricity is clean and environmentally friendly. In fact, his research indicates that flooding peatlands for Hydro-Quebec's mammoth Great Whale project could actually create as much greenhouse gas as electric utilities now produce by burning coal.

Although his message would undoubtedly have cheered critics of the giant utility's plans to flood a land mass almost the size of Prince Edward Island and sell the power to the northern US, the scientist planned to exercise scholarly caution in his address. While he advocates temporarily shelving the Great Whale project until he has more results from his study on peatlands, he also says that "at this point we don't know whether the flooding of peat is going to be a big problem."

To find the answer to that question, Rudd plans to continue his research at the Experimental Lakes Area (ELA), a unique preserve on Crown land in northwestern Ontario. After spring breakup, his team of researchers will flood peat that surrounds one of 46 pristine and protected lakes at the ELA site east of Kenora.

Courtesy of Barbara Robson

The team now has in place a low dam, a weir, a long dock that stretches far over the peat, and a handful of measurement devices embedded in it. In a peatland, Rudd explains, vegetation draws carbon dioxide from the atmosphere and stores that carbon as peat. In natural pools of water or man-made reservoirs, the process is reversed. The peat decomposes and gives off heat-trapping carbon dioxide and methane, an even more potent greenhouse gas. "What we're wondering is whether or not this may go on in reservoirs at a substantial rate," he says, adding that "we've seen it happening in one particular reservoir in northern Manitoba."

IF HE HAD BEEN ALLOWED TO ATTEND THE NEW YORK hearings, Rudd also planned to deliver this warning: "We should wait before we begin these large developments which are very hard to turn around once they're started....I'd like to see a pause (in the Great Whale development) until we know better the consequences." Not surprisingly, the refusal to allow his travel and speech raised strong suspicions. Had a government too sensitive to Quebec's economic interests found the political correctness of his science a tad wanting and tried to muzzle it?

In defence of its veto, a government official explained that the scientist's findings were preliminary, his theory too little tested to be of value to the US forum. But David Schindler, the country's foremost freshwater researcher, and the one most familiar with Rudd's work, disagrees.

In an interview on CBC-radio's science program *Quirks and Quarks,* Schindler countered, "He had plenty to say to that panel on the effects of the Great Whale project on fisheries and wetlands and the release of methane to the atmosphere. I view this as just the usual excuse of government bureaucrats to keep scientists from giving scientific information in forums that might interfere with the Canadian political agenda."

Schindler should know. The Rhodes scholar set the standard at the Experimental Lakes Area, a stunning pace of science that has changed politicians' minds and helped to transform environmental laws in several countries. He was research leader there two decades ago when Rudd did his doctoral research at the site, inquiring into methane, then known as swamp gas. In addition to finding that the potent greenhouse gas rises from peat flooded for reservoirs in northern Manitoba, the team discovered that very high rates of emissions of methane rise from natural pools in Canada's northern muskeg.

Schindler, a bullish and extraordinarily imaginative scientist, has had his own share of headaches in a federal bureaucracy that is as sensitive to politics as it is nurturing of science. After 20 years at the Experimental Lakes Area, he left for the University of Alberta. But during his stint at the remote, large-as-life laboratory, he brought immense recognition to Canada.

At the ELA, he demonstrated so persuasively that phosphates in detergents harm freshwater lakes and acid rain kills at unforeseen levels, that he influenced the laws of North America and Europe. The Stockholm Water Foundation, the Royal Swedish Academy of Sciences, the International Water Supply Association, and the International Association on Water Pollution Research and Control said as much last summer when jurists awarded Schindler the first Stockholm Water Prize, a $150,000 (US) award equal in status among freshwater scientists to a coveted Nobel.

"The Stockholm Water Prize to David was a very timely award," says Andrew Hamilton of the Ottawa-based Rawson Academy of Aquatic Sciences. "It signalled something that a lot of us have believed for a very long time — that the ELA program was in a class by itself."

HOW DO YOU TEST WHETHER A SPECIFIC HUMAN ACTIVITY — washing clothes in phosphate detergent or smelting ore without scrubbers on smokestacks — causes inordinate harm to the environment? Take a small piece of nature, pollute it with phosphorus or acid or other nasty stuff and test the water, the fish, the microbes and everything else for a while. The rare chance to do just that drew Schindler to the bureaucracy and held him there for two decades. The result: science which offered an immense weight of evidence; science which literally changed human behaviour.

But the road to high honour was often marred with bureaucratic bumps. For a time, Schindler was under unusual scrutiny by Canadian government officials. A US citizen, he was invited to speak of his research on acid rain's impact on lakes before US pollution control agencies and Congress. He wanted to advocate more stringent controls on pollution than the Canadian government was willing to set. Officials in the Department of External Affairs demanded copies of his speeches in advance.

The attempt to muffle his voice, he recalls, failed only because he gave up his effort to become a

Canadian citizen. The External Affairs Department couldn't touch him because censorship would have been an interference between a US citizen and the operation of his government.

For that reason, perhaps, Schindler has a great deal of sympathy with Rudd, or any other government scientist who is denied permission to say what he knows and clearly state what he suspects. "It is infuriating that we have some of the best scientific experts in the world inside the federal government and the public does not have access to what they know because they are used as political tools," he fumes.

❦

IN MID-OCTOBER, RUDD HIKED OVER BEDROCK AND THROUGH leafless bush en route to Lake 979, the lake whose boggy shoreline will be flooded. He stopped by the weir, unpacked his knapsack, and put on a pair of surgical gloves. His research partner, Carol Kelly, held open a ziplock bag as Rudd stepped into the icy water. On that grey, chilly morning, they wore surgical gloves and used teflon bottles and plastic sealers to avoid tainting the water samples with contaminants from their hands or their clothing.

That careful attention to detail, the development of scientific techniques, and the willingness to spend years, not just months, on research is what work at the ELA is all about and why so many institutions value it. Rudd's team includes a member of NASA's Mission to Spaceship Earth, who will use measurements of carbon dioxide emissions at the ELA reservoir to verify satellite data sensed in space. It also includes researchers from three US and six Canadian universities and two Canadian utilities — Ontario Hydro and Hydro-Quebec.

Ironically, the hydroelectric utility that might first be wounded by Rudd's research is one of his many supporters. Hydro-Quebec has donated $30,000 to another aspect of his work — the detection of mercury leached into man-made reservoirs. The 44-year-old scientist first studied mercury in water in the tragic pollution of the English-Wabigoon river system a few hundred kilometres east of the ELA campsite. Next he investigated it in northern Manitoba where Manitoba Hydro made its giant reservoirs.

"Hydro-Quebec is taking the same position as Ontario Hydro — that they'd sooner know now (about mercury) than find out later," says Rudd. To date, the two utilities have no financial interest in the research into gaseous emissions from northern reservoirs.

❦

THE RESEARCH PARTNERSHIPS WITH PEERS ALL OVER THE continent are just dandy for science on the scale of experiments at the ELAs. But these partners rarely pay for ELA maintenance. Instead, the Department of Fisheries and Oceans picks up a $500,000 yearly tab for research at the lakes and a ramshackle campsite at the end of an old logging road off the Trans-Canada Highway.

More than two decades ago, when the ELA was created, the now defunct Fisheries Research Board of Canada allowed some adventuresome people an adventure, as the Rawson Academy's Hamilton describes it "for those who wanted to be part of an undeclared and undefined conspiracy to save the planet." Their first nudge was the 1965 interim report of the International Joint Commission on pollution in the lower Great Lakes. To safeguard the lakes, to find out how to reverse their unhealthy state, and to protect other freshwater bodies from noxious pollutants, the simple notion of testing the waters in a smaller body of water was born. But it was born far beyond the narrow mandate of the Fisheries Department that now parents the ELA.

"Now within government the research must be defended on the basis of the departmental mandate," says Hamilton, who takes a special interest. He has been instructed by government to find a new base of future support for the ELA. Some think it an impossible task. Hamilton thinks it impossible that any government department would allow such a vital and highly regarded research program to wither while it looks for financial support from elsewhere.

Fortunately, the civil servant on loan to the Rawson Academy has a great deal of ammunition. The scientists' campsite may be spartan — from time to time there are leaky roofs on the labs, and the sleeping quarters, converted trailers, have only mattresses on the floor — but the pride among researchers who arrived for doctoral work and stayed on is exceedingly strong. The ELA is not only rich in natural resources, it has a wealth of influential researchers who have worked hard together for a dozen or more years.

Said Schindler at the 25th anniversary of the Freshwater Institute: "The history of the eutrophication section and especially ELA has been such that I've found the rest of the world is running around saying: 'Hey, how do we get ourselves an ELA.' This sort of approach is spreading like wildfire, but most people are amateurs. They don't have any idea how to do it as well as the ELA group does."

JUST OUTSIDE THE CAMPSITE, THE JACK PINES ARE REGENERATING. A lightning strike and forest fire in 1980 forced the scientists, the cook, the camp manager, and visiting scholars to head for the highway three times. "The last time they evacuated us, we got up the first hill and looked back and saw the smoke coming into the camp," recalls Ken Mills, the chief fish researcher who has worked at the camp since the early 1970s. "We thought it was gone. We thought that was the end."

With a turn of luck, the fire arrived at the camp at dusk. The winds dropped, the humidity climbed, and firefighters from the Ontario Ministry of Natural Resources saved the bareboned, precious bush camp. Another bit of luck saved Schindler in an earlier fire when his mission to rescue equipment from a lake ended in a plane crash. Tragically, the pilot died.

Camp lore, however, is more often preoccupied with what Schindler calls "another form of lightning." These are the serendipitous mergers of good science with pieces of evidence so graphic, so easily comprehended, that schoolchildren and politicians who have no interest in science couldn't fail to understand that something was amiss and have a notion of how to correct it.

The first was an aerial photograph taken in September, 1973, of two basins in one lake divided by a sea curtain — the far basin a soupy green from the thick algal blooms encouraged by phosphorus, carbon, and nitrogen. The near basin, which also received nitrogen and carbon, was normal by every appearance and measurement. That photograph, reproduced in 400 textbooks, had a firm impact on North American legislators who'd been told by soap manufacturers that phosphates had nothing to do with the tainted Great Lakes. They did not ban phosphates in detergents as many European nations did, but most jurisdictions set limits on phosphorous in sewage effluent entering the lakes. Public opinion did the rest.

"The improvement in Lake Erie has been dramatic," says Mills. "The walleye in that lake have started to come back. Where I grew up in Wisconsin you couldn't go swimming on the beaches of Lake Michigan because of the bacterial count. That isn't a problem any more. That's been eliminated. There really has been an impact of the program out here."

Schindler knows well what most impresses politicians. "Pictures themselves aren't good science," he says. "But it's often hard to put good science in a form that people can grasp easily. Twice in 20 years we were lucky to get something that was easy to put on a coloured slide."

The second strike was the publication in the journal *Science* of two photographs of lake trout alongside Schindler's paper on lake acidification. Mills took the first photograph of the healthy, fleshy trout in 1979, a few years after the team began trickling acid into Lake 223. The study spanned eight years before its findings were published. But just three years later, in 1982, Mills photographed a sickly trout that could no longer reproduce. It had wasted away in water no more acidic than water American policy-makers had tried to dub harmless.

> SCIENTISTS AT THE ELA DRAMATICALLY DOCUMENTED THE ENVIRONMENTAL EFFECTS OF ACID RAIN AND PHOSPHATE DETERGENTS.

Mills also showed the recovery of an acidified lake with a third picture of a fleshy lake trout. But he was chagrined to discover that Canadian diplomats at the Washington embassy had little regard for scientific protocol. The picture was plopped on a brochure that touted the Canadian–US acid rain accord before Mills had a chance to publish his findings or his third potent graphic in a peer-reviewed scientific journal. The right politics, it seems, can mean a speedy advancement of science.

THE MOST IMPORTANT DISCOVERY AT THE ELA, HOWEVER, may well be at one lake that is neither polluted nor manipulated, but just scientifically monitored. For more than two decades Schindler's team has recorded the ice-cover, winds, temperature, and other records of weather, hydrology, lake chemistry, and biology at Lake 239. The little lake gives a reference base for all other experiments. The 20-year record may also provide a preview of the greenhouse effect on a region that has been predicted to suffer the greatest effects of global warming in summer. Already the data show a 2°C rise in air and lake temperatures, a three-week increase in the ice-free season, stronger winds, less precipitation, and a concentration of most naturally occurring chemicals in both lakes and streams.

The political response to these findings was neither to muzzle them, nor to rush them to print. They were simply ignored, even after publication in *Science* in November, 1990. "In the past four years, not a dollar has been spent by the Canadian government on climate change related to fresh water," says Schindler. Understandably, he's annoyed.

Meanwhile, the long-term record is valued immensely in other countries. "There is no other experiment or project of this length and magnitude going on anywhere else in the world," says Bo Krantz, secretary general of the Stockholm Water Foundation. "It must be of infinite value to any limnologist to be able to draw on the data base from fullscale experiments that have been collected during almost a quarter of a century....The ELA has brought Canada to the forefront in water science."

Krantz once visited the campsite and last year wrote back that it would be unthinkable for Canada, and the world's scientists, to abandon the ELA research. "As I see it," he said, "the ELA is not a question of only national concern — it has wide international and, for that matter, intercontinental impact." His suggestion: Canada should turn the campsite and watersheds into an international preserve through the United Nations, or by inviting researchers from other countries to wage experiments in the remote lakes of northwestern Ontario.

The future health of the ELA research program and the fate of Great Whale remain to be seen. Rudd's science has the potential to prevent human folly in northern Quebec if politicians have the patience to wait for his results. Political leaders have the chance to affirm healthy research in northwestern Ontario if they have the courage to care for science, politics, and the environment. ■

Questions and Concerns

READING 5

1. The speaker, Tom McMillan, a former federal Minister of the Environment, starts this speech by praising environmental activists for their work on the acid rain issue. He also acknowledges the success of their campaign. But the thesis of his remarks seems to be critical. Why?

"The Real Story About Canada's Acid Rain Campaign

in the United States"

Notes for Remarks

by

The Honourable Tom McMillan

Canadian Consul General to New England

to

The National Conference

on

Government Relations

(Consultations and Public Policy)

Chateau Laurier Hotel

Ottawa, Canada

2 October, 1991

Courtesy of Tom McMillan

Let me begin by saying how glad I was to be invited to participate in this national conference on Government Relations.

It is a particular pleasure to share a platform again with the Honourable John Fraser, Michael Perley and Adele Hurley.

Back when John Fraser became Canada's Minister of the Environment, over a decade ago, acid rain was not a household term in Canada. He, more than any other Canadian politician at the time, understood the havoc acid rain was wreaking on Canada's natural and built environments. And, almost single-handedly, he caused it to be placed on the public agenda of the country. John continued to provide leadership on the issue long after he left the Environment portfolio. While we were in Opposition, he and I served together on the Parliamentary Sub-Committee on Acid Rain, which produced the landmark report <u>Still Waters</u>. I greatly valued his wise counsel when I was Minister a few years later.

If John Fraser was the acid rain pioneer in the political forum, Adele Hurley and Michael Perley were the two non-government leaders of the cause in Canada throughout the 1980s. I was Opposition Spokesman on the Environment from 1980 to 1983, and then served as Environment Minister from August 1985 to December 1988. From the perspective of both Opposition and government, I can say this without fear of contradiction: Michael and Adele were the pre-eminent environmental lobbyists in my time in politics -- articulate, knowledgeable, smooth, telegenic...public personalities in their own right. These two shrewd operators could hold their own with the sharpest of the American environmentalists, who, on the whole, are much more professional than our own.

I am, none the less, deeply concerned about the broader political culture of the country that compels Michael and Adele, and all other successful environmental lobbyists in Canada, to operate in a certain way that partly explains why this country is such a difficult one to keep united and almost impossible to govern. Let me elaborate with the candour that befits an off-the-record session like this one.

For the past two years, I have been living in the United States. From that vantage-point, I am struck by how fundamentally different Canada and the U.S. are in at least one respect. Ours is an extremely regional society, with deep and increasingly troublesome cleavages -- cultural, linguistic, and economic. Put simply, the country may lack the mortar required to keep it all in one piece.

The U.S. is also a highly regional country. And they, too, have deep cleavages. I am particularly shocked by the extremes of wealth and poverty. But, unlike Canadians, Americans are unified by potent myths and symbols that prevent their country

2

from blowing apart when a crisis strikes, whether it be the Civil War, Viet Nam or Watergate. In the U.S., the flag, the national anthem, the Constitution, the Presidency (if not the President himself) -- all these contribute to an over-arching American Idea that every American, native and immigrant alike, embraces. Patriotism is a civic religion in the U.S., and it binds Americans to their country and to one another with the force of a powerful drug.

By contrast, our national symbols and myths -- the flag, the monarchy, bilingualism, the Constitution and others -- tend to divide Canadians rather than unite us.

Our regional and cultural cleavages, which are deeper than those in the U.S. to begin with, are exacerbated by a political culture that thrives on conflict and confrontation. Indeed, we institutionalize this characteristic, for example, through our highly charged style of parliamentary debate. In Question Period, especially, success is measured not by how constructively or insightfully a debater approaches an issue but by how much blood he or she can draw from an opponent's jugular. The media reward success and punish failure by this one standard. Such does not apply to the same extent in Great Britain, where the Opposition is required to give Ministers advance notice about issues they intend to raise. The roster system for Ministers in Question Period has the same effect of civilizing parliamentary conflict in Britain.

In Canada, television, in particular, has rendered politics a blood sport. That medium constantly serves the public not information and analysis but contrived conflict. As often as not, this conflict takes the form of showcasing so-called public interest groups who know instinctively that the surest way to get a 30-second clip on the newscasts is to kick the hell out of an adversary -- usually the government. A reasoned and balanced response to a government measure is a prescription for being ignored by the media. Interest groups who know how to wield the sharp-edged dagger are the ones that become media stars. The rest perish. The extent to which the media and public interest groups stroke one another in Canada is so fraught with danger for the health of public discourse in this country that it puzzles me that more people are not alarmed. But, then, who in the media would report the alarm? No matter what the issue, thanks to interest groups, the media feed the public a steady diet of negativism that can't help but breed cynicism towards everything that smacks of politics.

Is it any wonder our politicians and political institutions are held in such low regard by the public? The reputation of Jesus Christ himself could not withstand the kind of abuse the media and interest groups, together, typically heap upon anyone who rises above the rank of water commissioner in the public life

life of this country.

Our country's politicians can hardly be expected to lead the nation effectively in these critical times when the phenomenon I have described robs them of the moral authority required for true leadership.

Lest anyone think I am merely rationalizing the problems the current Tory government in Ottawa is having with the public, let me stress that this phenomenon affects virtually all governments in Canada...a fact David Petterson and Bob Rae would be quick to acknowledge here in Ontario. Political honeymoons in this country are, as Thomas Hobbes would say, "poor, nasty, brutish and short."

I believe that environmental groups in Canada have done a great deal to educate the public about pollution and to encourage politicians to act against it. But, as former <u>Toronto Star</u> environmental reporter David Israelson wrote in his book, <u>Silent Earth</u>, "Too often many of Canada's environmentalists appear to be tough, but they're really just strident."

In the early days of the acid rain movement, the Coalition on Acid Rain was tough but rarely strident. And that approach produced results. But, ironically, as the cause gained momentum, **stridency replaced toughness**. Then, no matter what the Canadian government did in the eyes of the Coalition, it was not enough and never on time.

The Coalition urged the Prime Minister to sharpen his attacks on the U.S. Administration for failing to act on acid rain. When he did so, in a major speech to the Americas Society, the Coalition said such an approach would alienate the American public. The Coalition called on the Canadian government to launch an acid rain publicity campaign aimed at American tourists in Canada. When it did, the Coalition branded the effort as puffery. The Coalition criticized the government for failing to get New Brunswick and Nova Scotia to sign federal-provincial acid rain accords of the sort already signed by the other five polluting provinces. When the federal government finally did so, the agreements were dismissed as legally meaningless. And so on and so forth.

The Coalition plunged the "damned if you do, damned if you don't" syndrome to new depths. Because this practice is too frequently followed by other environmental groups as well, it is difficult for an Environment Minister to convince Cabinet colleagues to act against pollution on the grounds it will pay political dividends. Opponents in Cabinet can all too easily say, as they do, "we'll be attacked whether or not we act." Because that's usually the case. So, politicians are sorely tempted to invest their political capital where the payoff is greater. It's only human nature.

Believe me, I have discussed this problem with former environment Ministers like Clifford Lincoln of Quebec and Jim Bradley of Ontario, and they say the same problem plagued them in their respective Cabinets.

Even when the Mulroney government acted exactly as the Coalition had urged in a number of instances, the Coalition attacked it. This group kept shifting ground, so its scope for criticism was forever boundless.

For their part, the media normally do not require a public interest group to be consistent. The media could always rely on the Coalition's highly articulate spokespersons to provide searing criticisms of the government, and that's all that counted...Not penetrating analyses but plunging stilettos!

The problem in environmental terms was that, after a while, Cabinet Ministers, senior bureaucrats and diplomats alike simply discounted the Coalition as a reliable source of counsel, let alone as a partner, in the acid rain cause.

In fact, the successful strategy the federal government followed to achieve progress with the U.S. government on acid rain was the very one the Coalition repeatedly attacked as fundamentally flawed and, therefore, certain to fail. This fact has been totally ignored by the media, although <u>The Globe and Mail</u> published a first-rate article on the subject by former Canadian Ambassador to Washington Allan Gotlieb on March 12th, 1990. "For some time," said Gotlieb, "Canadian environmental lobbyists have been critical of everyone's efforts except their own. It's a good time to ask ourselves if this criticism is fair or misplaced." Clearly, the former Ambassador, like myself, believes the criticism <u>is</u> unfair.

In the case of the Coalition on Acid Rain, their basic premise was that, in the persons of Ronald Reagan and George Bush, the President of the United States would never act on acid rain. So, the federal government's efforts had to be focused on Congress, where genuine allies could be found or made. According to the Coalition, pursuing the President was not only a waste of time but also a dangerous diversion.

The coalition, of course, had a vested interest in such a Congressional focus because their lobbyists had some access to legislators and their staffs. They had no access to the President.

The collective wisdom in the Canadian government saw the situation very differently. It held that the President was pivotal in the U.S. political system. No matter how well-disposed Congress might be towards acid rain controls, progress was impossible as long as the President himself was opposed or indifferent. He had ways of influencing the agenda of Congress and how members voted. Ultimately, the President could also veto Congress on acid rain controls, and his veto would be sustained if, as seemed likely, Congress was at all divided.

That is exactly why, contrary to the Coalition's urging, Prime Minister Mulroney made acid rain the centrepiece of his bilateral summits with both Ronald Reagan and George Bush. I participated in one of the Mulroney-Reagan summits. And I made a presentation for the Canadian government when Ronald Reagan sent then Vice-President Bush to Ottawa to get, in Bush's words, an "earful" from the Canadian government on acid rain.

I pay tribute to the Canadian Coalition on Acid Rain for the superb job it did lobbying national and state legislators in the United States. The Coalition, along with American environmentalists, helped forge the consensus in Congress on the landmark acid rain amendments to the U.S. Clean Air Act passed last year.

But one fact has been ignored in all this: it was the President of the United States, George Bush, who, in June 1989, sent Congress the clean air bill on acid rain in the first place. And he put the considerable authority of the Oval Office behind the campaign in Congress to get the bill passed. The Prime Minister and his government -- including committed Canadian diplomats in Washington -- deserve, but hardly ever receive, credit for their dogged and skilful advocacy at the presidential level. George Mitchell, now the Senate Majority Leader, told me when I visited him in Washington as Environment Minister that the Canadian Prime Minister had been the only person able to get the U.S. Administration even to think about acid rain during the Reagan years. The groundwork laid with Reagan finally paid off in spades when Bush became President.

By the same token, the government was wise to ignore the Coalition's advice that it should not seek a Canada-U.S. Acid Rain Accord to enshrine each country's obligations to the other in this area. As Mr. Gotlieb pointed out, "the views of some of our lobbyists notwithstanding, Canada does not derive its rights from the U.S. Congress. It derives them from treaties duly entered into between sovereign governments."

My purpose here is not to detract from the achievements of the Canadian Coalition on Acid Rain. Let me repeat, I think the Coalition spokespersons were the pre-eminent environmental lobbyists in Canada in my time in politics. And the organizers of this conference were right to include them on the agenda as a model of success.

But, if the international lobbying campaign on acid rain was a spectacular success, as in fact it was, that success had many parents, and not just the Coalition. In the environmental field, as in life in general, only failure is an orphan.

It may well be that, in the early days of the environmental movement, activists had to be strident to get attention for their cause. But, now that environmental problems are at the top of the public agenda, it's time to mothball the dated ways and rhetoric of the past. The media should demand of public interest groups the same high standards of performance and integrity that those groups demand of the politicians they shadow. And the interest groups should demand no less of themselves. Having served in Cabinet, I know that such a constructive, non-confrontational approach would be a lot more influential as decision-makers seek answers to the weighty problems that, these days, threaten the very survival not only of our country but, indeed, of the planet.

Questions and Concerns

READING 6

1. Prevailing public opinion seems to be that "acid rain" has been beaten now that the new Clean Air Act is law in the United States; that acidic deposition has been declining steadily and will continue until it is no longer an environmental problem. The subject has all but dropped from view in the media. In the following reading, what leads you to believe that the battle is not yet over?

Signs of progress

Once the fundamental decision to reduce acidic emissions was accepted, some of Canada's largest producers of sulphur dioxide and nitrogen oxides showed commendable speed and effectiveness in implementing control measures.

Ontario Hydro's coal-fired generating stations currently emit about 15% of all acid gas emissions in Ontario, or 1% of the North American total. During the 1970s, the demand for electricity increased. So did the emission of acidic gases from the utility's plants, reaching a peak of 531 000 t in 1982. That was the year Ontario started regulating annual emission limits. By 1990, Ontario Hydro had cut its annual total emissions to 245 000 t, a drop of more than 50% in eight years. Continued progress is assured by the fact that the annual limit set by the province for the years 1990–93 is 280 000 t, and from 1994 on is 215 000 t.

Ontario Hydro describes its abatement strategy as a portfolio of demand-side and supply-side measures that allows the utility to implement the most appropriate measures where and when needed (D.B. Curtis, Ontario Hydro, personal communication). On the demand side, it has used information programs, incentives to improve efficiency, and other measures to influence the amount and timing of consumer demands for electricity. On the supply side, it has reduced its use of coal as an energy source, in favour of hydraulic and nuclear generating technologies. In existing coal-fired facilities, Ontario Hydro is using low-sulphur fuels and is committed to installing emission control equipment. It is also examining methods and technologies that will further diminish production of sulphur dioxide and nitrogen oxides.

Apart from thermoelectric power generation, the other major industrial process that produces acidic air pollution is the smelting of metals. Two of Canada's largest mining and metallurgical companies, Inco Limited and Noranda Inc., have vastly improved their pollution records in recent years.

The acknowledged giant among industrial producers of sulphur dioxide emissions has long been Inco Limited. In 1969, Inco's plant at Copper Cliff, near Sudbury, Ontario, was releasing sulphur dioxide into the atmosphere at a rate of 5 500 t/day, or about 2 million tonnes annually. Since then, a combination of positive company initiatives and government emission control regulations has reduced annual emissions to 685 000 t.

To meet future regulatory limits, Inco has developed programs that will lower its sulphur dioxide emissions to 265 000 t annually by 1994 and is studying the possibility of reducing emissions further in the years ahead (E. Kustan, Inco, personal communication).

In 1980, the Noranda Minerals copper smelter at Rouyn-Noranda, Quebec, was emitting just over 550 000 t of sulphur dioxide annually. In 1987, Noranda, the federal government, and the Province of Quebec signed an agreement whereby Noranda would reduce its sulphur dioxide emissions by 50% by 1990. Extraction of sulphuric acid, new smelting technologies, and the replacement of a portion of the high-sulphur ore concentrates by recycled metals have all contributed to a 90% reduction, by the year 2000, of sulphur dioxide emissions from the 1980 levels while conserving its established smelting capacity (D. Coffin, Noranda, personal communication).

Reproduced from: Government of Canada, 1991, *The State of Canada's Environment.*

Discussion for Acid Deposition

1. It is a fascinating notion that an environmental crisis like pollution causes symbolic or spiritual damage; that in the case of acid deposition, Canadians are demoralized by its attack on national symbols such as maple forests, the loon, and many heritage buildings. From personal experience we know this to be true. Pollution is depressing. Yet its effects on people are rarely expressed in these terms. Normally we talk about the economic costs or the physical damages. Discuss the emotionalism of environmental problems. How they affect you; your community; and the country at large. Should we pay attention to this side of the problem? Can this be considered a "cost?" Can major polluters be expected to curtail their pollution because of the emotional and spiritual damage they are causing?

2. There is a serious contraction between the valuable work being done by scientists at the Experimental Lakes Area in northern Ontario and the lack of willingness on the part of the federal government to support that work. Considering the calibre of the scholars involved, the many scientific papers they have published about their work at the ELA, the acknowledged importance of their results to the world scientific community, the influence their studies have had on phosphate and acidic pollution regulations, and the relative low cost of sustaining their operation, why would the Canadian government not support their research? What does this say about some governments and the depth of their commitment to improve and protect the environment? What should be done to ensure the ELA survives?

3. Former federal Minister of the Environment Tom McMillan says that the Canadian government and George Bush were not given enough credit for the passage of the *Clean Air Act*. He also claims that environmental activists working for the Acid Rain Coalition were never satisfied when politicians and governments responded to their demands; that they were too strident and lacked integrity. Discuss the role of environmental groups and how they influence government. (Also consider the hundreds of millions of dollars spent in North America by industry lobbyists to influence politicians on environmental issues.) Why is it many people look upon environmentalists as "fringe" or "radical" when they are trying to protect the environment?

4. Consider the relationship between the economy and the environment. In recessionary times, there is less pollution because their is less industrial production. But there is also less capital to invest in environmental protection, and the attention of the public is diverted from the environment to concerns such as unemployment and high taxes. If, in times of prosperity, there is more pollution and more money, what should be done to ensure the environment need not suffer through these economic cycles?

5. Legislation like the *Clean Air Act* in the United States requires industries by law to reduce their output of sulphur dioxide and nitrogen oxides. In Canada, no such legislation exists. So New Brunswick, for example, is reporting acidic deposits of over 12 kilograms-per-acre annually when it can only safely absorb around 3.3. Furthermore, the Canadian government agreed in 1985 to spend $150 million by 1994 for acid-pollution cleanup measures but, unbound by law, it had only spent $65 million by 1992. What does this suggest? Do we need stronger environmental protection laws?

Introduction

The Atlantic Cod Fishery

This issue examines the decline of Canada's Atlantic cod fishery. In 1992, a two-year moratorium was placed on fishing for the northern cod, one of the country's most important natural resources. This is a classic example of a "tragedy of the commons," where a common resource is shared by competing interests and inevitably ruined because no agreement can be reached on managing it sustainably.

Canada's valuable northern cod fishery off the coast of the Maritime provinces reached the point of exhaustion from overfishing in 1992. In response, the Canadian Department of Fisheries and Oceans imposed a two-year moratorium on cod fishing, putting about 20,000 fishermen and fish plant employees out of work. The cost to the country through aid and compensation paid out during the ban will be between $300 million and $600 million, in addition to the adverse economic and social "ripples" that emanate from the collapse of a key industry on which an entire region depends. (In 1990, the northern cod accounted for 381,819 tonnes or about 32% of the total catch for all species in the Atlantic fishery. Roughly 1,000 communities along the Atlantic coast are wholly or mostly dependent on the fishery.)

There is one obvious reason for this grave ecological event: Too many fish were taken from a resource that could no longer bear the exploitation.

Canadian government biologists are partly to blame. They consistently overestimated available stocks throughout the 1970s and 80s. This led the Department of Oceans and Fisheries to establish annual Total Allowable Catch (TAC) quotas that were higher than the cod population could bear. The TAC was set at roughly 20% of the "estimated exploitable biomass." As it turns out, biologists were operating with inaccurate information, using collection and sampling methods that overrepresented the real stocks. Fishing fleets may have actually been taking 45-50% of the total biomass (the weight of all cod in the system).

This coincided with an increase in harvesting capability and a lack of restraint on the part of the domestic fishing fleet. In the early 1980s a small inshore "dragger" ship could net about 23 tonnes of fish. It's equivalent today can take 164 tonnes. According to the Scotia-Fundy Groundfish Task Force, a federal task force formed in 1989 to investigate declining yields, as the fishing fleet became equipped with larger and more sophisticated ships and gear during the 1980s, groundfisheries (cod is known as a "groundfish") off the coast of Nova Scotia were badly harmed by overfishing, misreporting of catches, and inadequate monitoring, surveillance, and enforcement by officials.

Canada's two large fish processing companies – National Sea Products and Fisheries Products International – added to the problem with huge offshore factory-freezer trawlers. These highly mechanized ships use radar to pinpoint the location of large schools of fish then catch them with bottom-dragging nets that scour the ocean floor. This indelicate method damages the marine habitat and pulls huge volumes of unwanted species – known as "bycatch" – along with the preferred catch. Trawlers looking for shrimp, for example, will commonly catch nine pounds of other species, including baby cod, for each pound of shrimp they land. The bycatch is thrown back into sea, dead. It is believed that as many fish are dumped along Canada's eastern seaboard as are landed.

The Canadian government blames the fishery's collapse on foreign trawlers operating at the edge of Canada's 200-mile sovereign fishing limit. While most of the rich continental shelf breeding ground known as the Grand Banks lies within the limit, the "Nose" and "Tail," two important migratory areas, extend outside the boundary and are fished heavily by foreign ships. The Northwest Atlantic Fishery Organization (NAFO), an international body formed to manage fishing in international waters of the Atlantic, has set harvest quotas for these sensitive "straddling" areas at the rim of Canada's limit. But the European Community (EC) has consistently objected to and ignored the quotas since 1986 when Spain and Portugal joined the EC. Trawlers from Spain and Portugal, persistent harvesters in the region, often take immature fish, hampering the regenerative abilities of the cod stocks. They are also sighted regularly fishing illegally within the 200-mile limit.

The depletion of the cod stock off Atlantic Canada is what American biologist Garrett Hardin calls "the tragedy of the commons" – the ruinous exploitation of a common resource such as cod (or clean air, clean water, the ozone layer) when it is shared by competing parties, each pursuing its own interests without limit. In the case of cod on the Grand Banks, domestic fishermen with large investments in equipment are competing with each other for dwindling stocks. They also know that if they don't take the fish a foreign vessel a few kilometres away will. Likewise, foreign fleets take as many fish as possible before there are no more to take. The inevitable result is ruin.

"At the center of efforts to maintain the world's fish supplies," writes Hal Kane of Worldwatch Institute, the environmental monitoring organization in Washington, "is the issue of property rights. Since fishermen can't own a portion of the oceans they often have no understanding of their stake in preserving what swims beneath the surface. Rather, they're inclined to take as much fish as possible – while it's still there to be taken."

The alternative to this headlong rush toward collapse is "sustainable yield," where a resource is managed to balance the breeding capacity of the fish with the harvesting capacity of the fishing fleets. It has worked with haddock, a species that was in serious decline several years ago in the waters between Nova Scotia and Cape Cod. In 1985, the International Court of Justice established a boundary that delineated portions of the fishery that would be controlled by Canada and the United States. Canadian fishermen were forced to operate under federal quotas in their territory. American fishermen had no quotas. The result: On the Canadian side of the boundary, the haddock are now more plentiful but larger. On the U.S. side, they have continued to dwindle.

Commercial fisheries throughout the world are experiencing similar declines. The Food and Agriculture Organization of the United Nations (FAO) has estimated that the world's fishing fleets cannot catch more than 100 million tonnes of fish annually without critically depleting stocks. That point was reached in 1989 according to Worldwatch. It is now estimated that about 90% of the world's commercial fisheries are overexploited. Of the 280 species monitored by the FAO, only 25 are under-exploited.

In the end, it's a matter of what one former fisherman says is a dire need to "build a new relationship between ourselves and the natural world," remembering the basic ecological principle of interconnectedness. If you overharvest, you ruin the resource and lose your livelihood.

"If the cod stocks are exhausted," writes Don Gillmor in Equinox, "it will be with the usual combination of complacence, willful stupidity and lack of political will that has undermined us in the past."

Questions and Concerns

READINGS 1 THROUGH 4

1. Cod were once so plentiful on the Grand Banks that ships literally ran aground on schools of fish. Yet cod are also a "fragile resource" because of the way they reproduce. As you are reading, consider the biological odds against the cod.

2. Think about the various groups fishing for northern cod and why they are blaming each other for the collapse of the resource.

3. Canada is spending millions to identify and apprehend foreign fishing vessels that are illegally fishing within its 200-mile protected limit. Do you think this money is being spent wisely? Does this vigilance seem to be paying off? What are the alternatives?

4. If foreign fishing fleets are ignoring quotas set by Northwest Atlantic Fishery Organization (NAFO) to protect cod stocks, what does this say about the authority of international organizations? What does it say about our ability to protect the global environment? Why would European nations act in their own interest if an important resource like the Grand Banks cod fishery is in peril? Ultimately, is this not against their interests too?

5. John Crosbie, the Minister of Fisheries and Oceans, mentions "unilateral actions" which Canada might have to take to protect the cod fishery. What do you think he means?

A FINE KETTLE OF FISH

Newfoundland, say the old-timers, *is* cod. So what happens when the cod suddenly go missing?

Article by Don Gillmor

In Petty Harbour, Newfoundland, just south of St. John's, 15 local fishermen in baseball jackets and beer-company caps pull out chairs around a wooden table on an early-February morning. They are men who make their living catching cod near shore with traps and baited hooks, and they are of the opinion that the Atlantic cod stocks have been fished to disaster. The problem, explains one, is the offshore fishery, the trawlers that drag the spawning grounds, sweeping up everything in their paths. Before long, the fishermen are all talking at once, the syllables formed at the back of their throats coming out in rolling Gaelic noises. Their voices overlap in keening outrage.

"Scointists say dragging don't interfere with spawning, but common sense will tell you otherwise."

"You can't fish the spawning grounds forever. When they're spawning, they're stupid, same as you and me. You can pick trout out of a river with your hands when they're spawning."

"Even a goddamned idjit knows he eats all the eggs, there won't be no chicken, for Jesus' sake."

The voices move higher, spilling out at an auctioneer's pace. A former draggerman who has worked the trawlers offers a partial defence of trawling, but he is summarily dismissed. "Since when did draggermen start telling the truth?" asks a neighbour. "They keep the three-footers, dump the small ones, bye."

One of the fishermen tells me how the trawlers misreport their catches; how hundreds of tonnes of cod are dumped at sea each year; how the fish are getting smaller, a warning sign of overfishing. "The last of the big fish is out there now," he concludes. "The technology wipes them out." The chairman brings the meeting back to order after pleading and hammering the gavel. Half a dozen men crouch, arms stretched in front of them, hands spread apart, an interrogative lift to their eyebrows. The fisherman beside me leans over: "It's a complete slaughter, right?"

In other parts of Newfoundland, down the Burin Peninsula, along the east coast, opinions vary widely, but cod remains the central topic of discussion. For the better part of 400 years, Atlantic cod have been the pride of Newfoundland, its lifeblood and saving grace. Since the first European sailing ships arrived, fishermen have praised its waters as the world's richest hunting ground for cod, a maritime treasure as inexhaustible as the primeval rainforests of British Columbia or the black topsoil of the Saskatchewan prairie. But such confidence has led to folly; many biologists believe that northern cod stocks are in peril. While offshore trawlers are bringing in good hauls, the inshore fishery is on the brink of disaster. As a result, Canadian scientists have retracted glowing forecasts penned for the fishery just two years earlier, replacing them with dire warnings. To avert ecological disaster, the federal government has slashed the Total Allowable Catch (TAC) by more than one-quarter, from 266,000 tonnes in 1988 to 197,000 tonnes in 1990. In Newfoundland, the news has triggered a devastating domino effect — plant closures, layoffs and the threat of collapse for entire villages.

Whether the stocks can now rebound remains in question, for Atlantic cod, *Gadus morhua,* are a fragile resource. Mature female cod lay more than one million eggs each along offshore spawning grounds, but few survive. Fertilized by rogue sperm, the eggs are at the mercy of tides, currents, seabirds, small fish and pollution; the slightest disturbance of water or wind can threaten an entire generation. Those which do survive must stay close to the sea bottom and migrate in schools from the deep water offshore to the shallows, evading humans and seals for seven years before attaining reproductive age. On average, only one egg in a million reaches maturity.

While fishery experts worry about rebuilding the ragged resource, there is little agreement on the causes of the current crisis. Across Newfoundland, a certain symmetry characterizes the debate. The offshore trawler captains do not think there is a problem, except for seals that eat the inshore cod; the inshore fishermen believe the offshore fishermen are the problem; the plant workers think it is foreign overfishing; the foreign experts say it is Canadian mismanagement; the provincial government blames the federal government; the federal government takes the visionary position that too many fishermen are chasing too few fish; an independent study suggests a serious resource problem exists, one that stems partly from inaccurate Department of Fisheries and Oceans (DFO) estimates of the cod stocks for the past 13 years; the scientists who made the estimates deny there is a resource problem; and a woman at the Petro-Canada station in Goobies, Newfoundland, believes it is all God's will.

At the heart of the problem is what American biologist Garrett Hardin calls "the tragedy of the commons." When a common resource such as cod is shared by competing parties, each pursues his or her own interest without limit. To reduce the catch seems folly to a fisherman, for the cod will only be caught a kilometre away by a competitor. In such a system, ruin of the resource is inevitable. Unless common good can replace individual need, fishermen will argue among themselves, communities will fight one another for survival and countries will plunder each other's stocks. Unless some solution can be found soon, northern cod stocks will be fished to exhaustion by the very people who depend on them most.

One of the oldest, most dependent fishing communities in North America is Grand Bank, a town of clean-lined maritime houses and 3,750 residents on the interior coast of the Burin Peninsula. At the centre of town on a winter morning, the *Grand Baron* is pulled up alongside the Fishery Products International Limited (FPI) plant, ice edging its gunwales, listing to one side as its hold is off-loaded. Even by the generous margins of marine aesthetics, Atlantic cod could hardly be called a handsome fish, with its small goatee, greyish pallor and slack, brainless mouth. Sixty centimetres long and thick-bodied, the fish slide along a steel chute, the split lengths wriggling downward as if alive. In the plant, dozens of workers stand at separate stations, trimming the fish and separating the flesh into fillets, nuggets, end pieces that are processed into blocks and sold for recutting into fish sticks. Machines box the fish, water flows on the floor, the tonnes of ice melting, spray hoses splashing. Boxes of cod tongues sit nakedly on a steel table, waiting to be packaged for the adventurous. In all, 71 percent of FPI's cod will be shipped to the United States, 12 percent to Europe and Japan and the remaining 17 percent to domestic markets.

Grand Bank is a company town, an FPI town. "There is no other industry," says Mayor Rex Matthews in the Town Hall boardroom. "Everyone is connected to fishing. The town is 100 percent dependent." Eight years ago, he explains, 14 trawlers worked out of the town and the plant stayed open year-round. But with the decline of the fishery, FPI has cut back operations: in 1990, it opened for only 20 weeks. Like other part-time plants, it is known as a stamp factory, a reference to unemployment insurance stamps, a minimum of 14 weeks of work per year needed to qualify for benefits.

On the radio, the weather report, cold with little relief expected, is sponsored by FPI — "Fishery Products International, building on the proud heritage of the past to ensure the success of the future." The successful future includes the closing of the Grand Bank fish plant and two others on the coast and the loss of 1,339 jobs. The proud past stretches back seven years, although only four could be considered genuinely proud. FPI was formed in 1983 from the assets of five financially troubled fish companies. The companies had undergone rapid expansion when Canada extended its fishing boundary to 200 miles offshore in 1977 but were soon done in by rising interest rates and soft markets. By the mid-1980s, the Atlantic Provinces' insolvent fish firms had been restructured into two corporate monoliths, FPI and National Sea Products Ltd. Both soon announced layoffs and closures. When the Grand Bank plant ran year-round, FPI's chairman, Vic Young, was a popular person in town, "the nicest man you'd want to meet." Today, it is hard to find anyone with a kind word for him.

Throughout Grand Bank's 303-year history, the cod and flounder fishery has directed the ups and downs of town life. In the 17th century, fishermen chose the site for its harbour, its pebbled beach and the abundant cod in the area. By 1693, the colony census showed one family, the Bournys, six servants and two other French settlers, Chevallier and Grandin. Eventually, the town turned English, then drunken, alcohol plunging the population "headlong into every species of brutality." A stretch of Methodism sobered the town, with dancing declared the devil's tool and card playing a sin. The two extremes united briefly in the form of the local schoolteacher, a lapsed priest and blackout drunk who lay on the floor of the schoolroom and lectured the class on temperance and the Christian path.

Through damnation and redemption, Grand Bank relied on cod. In the early years, men headed out in shallops, 9-to-12-metre boats that accommodated up to five men fishing over the side with hooks and lines. By the late 19th century, however, the fleet consisted of larger, more seaworthy schooners capable of carrying men farther out onto the Grand Banks, the rich offshore spawning ground southeast of Newfoundland. Most had adopted trawls, dragging lines along the bottom. But even then, some fishermen worried that dragging would catch female fish full of spawn, ruining the fishery. In 1863, Captain Hamilton of the British Navy reported that the new gear was, in fact, landing large numbers of roe-laden females: "I believe our own men acknowledge the impolicy of it but plead necessity." The British government was petitioned to stop trawling on the Grand Banks, but no action was taken, and many of those opposed to the new technology finally adopted it, forced to compete. The result was a widening rift between the men who fished close to shore and those who worked the trawlers, a division that has stood for 120 years with the same quotidian frictions and complaints.

Ashore, the cod were cured, salted and laid to dry in the sun on the pebbled beach of Fortune Bay. While the men were at sea, the women processed the fish, stepping barefoot and bonneted among thousands of flayed white codfish curling like paper at the edges. Families existed largely on credit, settling with the British merchant after the summer and taking more supplies on credit for the winter. It was a marginal, hard-won existence; a year's work might clear $20 after the merchant was paid. To compensate, most embraced rum, religion and the Protestant work ethic and accepted a widespread fatalism, a belief that their time would come when it came. It was not unfounded. Over the past 128 years, 113 Grand Bank vessels and hundreds of its men have been lost to the sea.

As early as 1885, Reverend Moses Harvey, a naturalist and theologian, talked of Newfoundland's "exhausted seas." The catches had been small and erratic, and Harvey feared for the future of the resource. Although the stocks recovered in time for World War I, the market took a dive in 1921, and the demand for cod remained low throughout the 1930s. The economy of Newfoundland began to unravel as the effects reverberated among retailers and others dependent on fish income.

A new crisis arrived in the 1950s with the salt-cod market in swift decline, edged out by the demand for fresh fish and by new technology. Powerful trawlers employing nets were rendering Grand Bank's schooners ineffectual and dated, their catches picayune and labour-intensive. Eager to embrace the changes, Grand Bank officials began lobbying for a fish plant. In 1955, after five years of negotiation, the plant opened with a fleet of three company-owned trawlers for the offshore fishery.

Grand Bank became a trawler town, its fortunes tied to the vagaries of the offshore cod fishery.

For 35 years, the trawlers and the fish plant have kept the community afloat, and some residents see little to fear in the current cod controversy. "There's more fish than ever," says former trawler captain Allister Stone, sitting back in his chair in the Town Hall boardroom, staring upward. "Course, I'm clear of it now. But more codfish. To me, there's lots of fish, and we're leaving them to the foreigners. I still thinks you don't have to cut them down much. Run through cod, 15 minutes on the sounder." Despite the perceived abundance, however, local inshore fishermen cannot make their quotas and have turned to flounder and lumpfish roe. Farther up the coast, the inshore cod fishery had failed utterly.

"Seals are eatin' them up, right?" says Stone. "Seals is driving them off the land. The inshore boats don't have no fish to catch."

Exactly how the Atlantic cod fishery reached its present precarious state remains a riddle, an ecological puzzle of humbling magnitude. An aquasystem is a difficult thing to manage; researchers lack information on the basic science and biology of the stocks most critical to commerce. They cannot see them, count them, predict their movements or prevent national and international fleets from fishing them piratically. It is an unsolvable equation in which all of the variables are unknown, and some of them are shifty and bloody-minded as well.

Newfoundland's government and its premier, Clyde Wells, have little sympathy for the researcher's problems. Wells, an avid sailor and the grandson of a fisherman, has paid keen attention to shifting fortunes in the cod fishery over the years. "Ten years ago, the scientists were saying, on the basis of the management system they were putting in place then, they would expect that by 1991 or 1992, the TAC would be 400,000 tonnes," he says. "As a result, Newfoundland fishermen and fish-plant processors built plants and bought new boats to fish what everybody thought was going to be a growing TAC. And the TAC did grow: in 1987, it was up to 256,000 tonnes, and in the fall of 1987, the federal scientists were advising that it should be increased to 293,000."

But the TAC was not increased to 293,000 tonnes. "The Newfoundland fishermen, who had been experiencing increasing difficulty with the inshore fishery, with fewer fish requiring greater fishing effort, knew that something was going wrong," says Wells. "They blamed it on increased pressure from the offshore fishery. They objected strenuously, and so did the Newfoundland government of the day." Federal scientists waded back through their research. According to their 1989 estimates, the cod stocks appeared to have tripled since the late 1970s. But the base population of the late 1970s had been greatly depressed by foreign overfishing. In addition, 1987 data indicated that the stock had grown by as much as 500 percent, a figure which proved to be woefully optimistic. Federal fishery managers' data suggested a TAC of 125,000 tonnes for 1990; the minister compromised at 197,000 tonnes.

The provincial government argues that federal scientists erred and that on the basis of their faulty analysis, the province built up its cod fishery and is now left with the consequences: widespread unemployment and, in the smaller villages, social ruin. "We can't say this is a God-caused disaster," concludes Wells. "This is a federal-government-caused disaster. They are the ones who have the exclusive jurisdiction to manage the fisheries; they are the ones who made the management decisions that resulted in this."

In the pagodalike octagons in the White Hills on the outskirts of St. John's, there is a different point of view. The octagons were built 10 years ago, when DFO officials decided to shift their operations from Water Street, down by the docks, to the edge of town, where they had more space and access to unpolluted waters. The DFO buildings are labyrinthine and inscrutable, hives of scientists and managers in small cells, a map of Newfoundland on every wall. Today, the fishermen do not get up to the White Hills much, and the scientists do not come down to the water.

"The industry is in a state of crisis," says Jake Rice, head of DFO's ground-fish division, "because all the people who want to make a living from it can't make a living from it. The stock isn't in a state of crisis." It is an economic rather than a resource problem. "There was no mistake made," says Rice. "In retrospect, the quotas shouldn't have been as high. But to say we made a mistake implies that we had the right data and interpreted it wrong."

Few people have ever implied that the DFO had the right data. The indices of abundance, the methods used to calculate fish populations, come from just two sources. The first is a sampling of commercial fish catches, based on the report of fish-plant technicians, offshore fishermen and federal government observers. The second is a government survey conducted every fall, when federal scientists take random trawl samples, calculating the number, age and weight of the fish and extrapolating from that data — a "Gallup poll" of the Grand Banks.

After collecting the data, federal scientists estimate the size of the stocks and the total biomass, or weight, of all cod in the system. Generally, the minister sets the TAC at a level roughly equivalent to 20 percent of the estimated exploitable biomass, a figure that should offer fishermen an optimal economic return for their efforts and give the fish stocks a base from which to grow. At the root of the current problem, say critics, is the fact that Atlantic fishermen have not been landing 20 percent of the biomass, as they thought. They have been taking 45 percent, possibly 55 percent. On the basis of unreliable data, Canadian officials have been making some dangerous decisions.

Leslie Harris, president of Memorial University of Newfoundland, in St. John's, headed an independent review panel commissioned in 1989 by the federal government to investigate the state of northern cod. Harris has a pipe, a fleeting smile and a Holmesian air of deduction. He relights his pipe and examines the methods that the DFO uses to estimate the number of cod. "All fishermen are liars, except for you and me," he says with a smile. "And I'm not too sure about you." It is a popular aphorism, one found on small plaques in tourist shops, but in the case of commercial fishermen, it contains an element of truth. "Fishermen, if they are fishing with quotas, will beat the system by misreporting," says Harris. "What is most likely misreported are fish not for commercial sale, which are simply dumped at sea and never reported." Cod fishermen will sometimes dump the smallest of their catches, keeping only the large fish. The fish are usually dead by the time they are thrown back in, but even the few survivors quickly die. When hauled from deep water, their swim bladders inflate rapidly, rendering them too buoyant to return to their sea-bottom habitat.

It is difficult to estimate how many fish are lost through dumping, but the numbers are significant. "Other fisheries, such as shrimp, take in substantial numbers of usually quite young cod," says Harris, "because cod feed on shrimp and the populations get mixed up during feeding cycles. And if your gear technology is indiscriminate, there will be cod in the shrimp catch. But the shrimpers won't save cod, they'll just dump it overboard. We had shrimp captains who told us that they dumped 10 tonnes of cod for every tonne of shrimp they took." The commercial data is biased, suspect; it is a fish story.

DFO scientists also use catch-per-unit-of-effort data as an indication of abundance, measured by the success of individual trawls on commercial boats. If it takes more effort — more time, more fuel, more men — to catch the same number of fish as in a previous year, the ratio suggests that there are fewer fish. But just the opposite has been happening in the offshore fishery, a sign, thought DFO scientists, that the stock was growing. As Harris points out, however, the improving catch ratios could just as easily have come from a better-equipped fishing fleet. "The quality of the net, the quality of the materials, the power of the vessel, of the winches, the configuration of the net — all these have changed drastically, become more efficient," he explains. Captains have become more experienced, electronic gear more sophisticated. "Killing efficiencies, tracking efficiencies, finding efficiencies have improved. There is no wasted effort."

Harris also takes issue with the DFO's heavy reliance on data collected during the commercial fishing season. "The fishing of the commercial fleet is done solely during the period of fish concentrations, of fish spawning," he says. "Fishermen don't try to catch fish when they're scattered all over God's farm. They catch them when they're all together." It is difficult to tell, then, whether the population is declining or not; the net will always be full, until the last fish is taken out.

Other inaccuracies could enter the picture when DFO scientists convert estimates of cod populations into statistics on biomass, the basis of the industry quotas. The conversion from numbers of cod to tonnage is made by determining the age of a sample of fish (the ear stone [otolith] of the cod has growth rings that can be counted like those of a tree). The average weight of the fish at a specific age is first calculated and then multiplied by the relative proportion of that age group in the system. Given the millions of fish in the equation, a miscalculation of even a hundred grams can have a pronounced effect. Complicating the issue is the fact that cod grow at different rates in different water temperatures: a 5-year-old cod from the southern coast will be larger than a fish of the same age caught off Labrador.

Another tricky variable is the annual survival rate of the eggs and larvae. Some years, cod eggs will drift into waters short on food but long on predators, and a poor "year class" will be born. A year class is similar to a school graduation class: some students are distinguished future captains of industry and Nobel laureates; others are lacklustre and dull, headed for daytime television. When a poor year class is recruited into the commercial fishery, roughly at age 3 or 4, the result is fewer fish — a disappointing fishery until a good year class comes along to bolster the ranks. In theory, cod should not be fished until they reach the age of 7, the age at which they begin to reproduce. But experts report that of the cod caught and aged, few were over the age of 8; the majority were between 4 and 7. We are, in essence, eating away the biological capital of the cod stock, no longer living off the interest.

The whole business of estimating cod is mathematically sound, carried out by diligent scientists and subjected to peer review, and it is as reliable, finally, as reading entrails. At each stage of the calculation, there is opportunity for wild mistakes. As the steps accumulate, the margin of error can lengthen, yielding an encouraging and unrealistic figure. "The biggest problem with the cod estimate," concludes David Schneider, a marine biologist at Memorial University of Newfoundland, "is that there is as much as a 50 percent margin of error."

The fundamental problems of managing fish stocks are aggravated by factionalism. During other recent crises — foreign overfishing in the 1970s, for example — the varied interest groups set aside their rivalries, banding together. Inshore and offshore fishermen, plant workers, scientists and politicians all agreed that they faced a common problem in the foreign rape of the continental shelf. They were a single, effective voice. Since then, the groups have drifted apart, establishing their own agendas, hiring their own lawyers. The Newfoundland Inshore Fisheries Association is suing the federal ministers of the Environment and of Fisheries and Oceans, claiming that they have fallen down on the job, that the fishery should receive the same environmental protection as do other natural resources.

As a political force, the factions are diluted and contrary, each proposing solutions that will anger another group. The response from the federal government has been timid and vague. The most entertaining solution has been to call on Newfoundland to diversify its economy, to branch out of fish. The province has a rich and pixilated history of diversification. Former Premier Joey Smallwood recognized the precarious singularity of the provincial economy, the narrow hope of cod. Plans were made to promote tourism, to duplicate the town of Rothenburg, Germany, and to line the first 32 kilometres of the Trans-Canada Highway east of Port aux Basques with flowering trees. There have been greenhouses, rubber plants, chocolate factories, failed pulp-and-paper mills, aborted hydroelectric projects and the stirring vision of the Flying L Ranch Company. The Flying L was started near Grand Bank in 1964 with 1,000 Hereford cattle from Saskatchewan. They died in the bog, poisoned themselves on marsh grass, were rustled nightly and starved in numbers, their feed blown across the island like dust. Within four years, the ranch was liquidated, the losses heavy. The most lasting diversification has been into unemployment, which includes between 18.3 and 23 percent of the province's work force, depending on whose figures are used.

Cod remains the cornerstone of Newfoundland's economy. Newfoundland *is* cod, a precarious resource; when fish stocks collapse, they tend to do so quickly, disappearing with the blithe speed of a stock-market run. If each fisherman were to own a finite, controllable part of the ocean — an impossibility — he would control his catch to obtain maximum benefit over a prolonged period. But fishermen vie for the same resource in the same space, and in such a system, there is no common good, only individual need.

"Fishermen can't restrain themselves because they are competing," says Schneider. "There are examples when this has driven a fishery into extinction, such as the collapse of the sardine catch off California. The collapse was dramatic, failing in three months." Schneider cites a combination of heavy fishing and a natural event, possibly a climate-induced variation.

From 1944 to 1945, the fishing fleet took 557,000 tonnes of sardines from the waters off California. On September 25, 1945, two fishing records were broken: 8,000 tonnes of sardines were received at a single port in a day, and one boat caught 243 tonnes. But in October and November, landings declined profoundly. By mid-December, no fish could be found off central California; the sardine population had been devastated. Because of the schooling tendencies of sardines, a survival mechanism, the landings had remained encouraging right up until the last fish was taken.

Ground fish such as cod and haddock are more resilient than sardines, requiring more effort, a greater act of will, to deplete. But we have proved up to the task. In Norway, fishermen decimated cod, herring and capelin stocks, causing the starvation of harp seals and the permanent migration of seabirds. In Canada, overfishing has also led to disaster. In 1955, the haddock fishery off Newfoundland's coast reached a peak, with landings of 104 million kilograms by Soviet, Spanish and Canadian trawlers. By the early 1960s, the haddock stocks had become commercially unviable and have remained so.

In 1977, the federal government extended its fisheries jurisdiction from 12 to 200 miles offshore, a decision, it said, that was largely prompted by a resource crisis. "The main thrust of the new 200-mile management regime," wrote officials, "is to rebuild the resource so as to provide increased catches and catch rates for Canadian fishermen." Management regime is perhaps too strong a phrase; we have little real control over the area, little management. Foreign trawlers violate the 200-mile line, largely immune from the two DFO boats that police the hundreds of thousands of square kilometres. There is a quota for foreign vessels, reached by agreement between Canada and the European Community, but it is openly flouted. Between 1985 and 1988, European nations were allocated a total of 35,000 tonnes of northern cod; they reportedly took 165,000 tonnes, nearly 400 percent above their quota. Even if Canadian jurisdiction were properly observed, the Grand Banks would be in danger; the nose and the tail, key spawning areas, just outside Canadian jurisdiction and are preyed upon by foreign trawlers.

In a fishery where such ruthless competition prevails and where we have little chance, really, of guessing the number of remaining cod, it is disconcerting to hear the certainty in the voices of those who state there is no danger, no crisis. There is a hint of smugness, a patronizing quality, a vast knowledge of the sea born of unfamiliarity. There is little room for empiricism, endless room for experts.

The cod stocks are resilient, a fact often pointed to by government scientists. But they are not nearly as resilient as we are. If the cod stocks disappear, a distinct possibility, consumers will eat turbot or redfish or halibut. The fishermen and workers will find other work, go on unemployment, welfare; many will move. Their lives will be diminished, dislocated and impoverished, but most will survive. We have made a habit of adapting — to bad air, mercury poisoning, suspect water, assorted chemicals. We have learned to coexist with toxic waste, acid rain and low-level radiation. Our ability to adapt to our own ecological vandalism is not necessarily a strong point. If the cod stocks are exhausted, it will be with the usual combination of complacence, willful stupidity and lack of political will that has undermined us in the past. "Ruin is the destination toward which all men rush," writes biologist Garrett Hardin, "each pursuing his own best interest in a society that believes in the freedom of the commons."

At the inshore fishermen's meeting at Petty Harbour, a man describes the endless rivalry of the sea: "If I find some fish, someone on the radio say, 'Are there fish there?' I say, 'No.' They'd be on me, boy. Fish'd be gone." In a video made by the Petty Harbour Fishermen's Cooperative, a half-dozen small boats bob in the bay, fishermen in brightly coloured rubber overalls stand and throw out hand lines and quickly haul them back, a 60-centimetre cod twisting dumbly at the end every time. But it does not end there. Farther offshore, a dozen trawlers move around one another like chess pieces, scooping out tonnes of cod with every landing, saving them from a neighbour's boat. ■

Net losses

The sorry state of our Atlantic fishery

By Silver Donald Cameron

AFTER A FEW rich seasons, the East Coast fishery is in deep trouble again — for the third time in 15 years. Like the ocean ecology that shapes it, the fishery is enormously complex, but the essence of this sudden new crisis is simple: our sophisticated fishing vessels are catching fish far more efficiently than anyone realized.

"Our technology has outstripped our science," says Dr. Leslie Harris, chairman of the federally appointed Review Panel on Northern Cod and president of Newfoundland's Memorial University. "We have underestimated our own capacity to find, to pursue, and to kill."

Trouble in the fishing industry is big trouble for Atlantic Canada, with its 65,000 fishermen and more than 40,000 fish plant workers. Fishing provides 10 percent of the region's jobs; as much as 25 percent in parts of Newfoundland. In total, it is a $1.5-billion business.

The fishery takes place in four main areas: the Scotian Shelf, stretching from the mouth of the Bay of Fundy to the northern tip of Cape Breton Island; the Gulf of St. Lawrence; the Grand Banks of Newfoundland; and the Labrador coast. The situation varies from one area to another, but nowhere in that vast expanse of ocean are the fish populations really healthy.

The two areas that dominate the news are the Scotian Shelf and the banks off Labrador — and the two are very different. The Scotian Shelf is within easy reach of inshore fishermen all along the coast of Nova Scotia and New Brunswick, and they fish a wide variety of species: cod, haddock, flounder, pollock, hake, herring, redfish, crab, scallop, lobster and others.

The Labrador fishery, by contrast, covers a vast area of the ocean east of the Labrador coast and north and east of Newfoundland, and its fishery is dominated by the stock known as northern cod. This stock is fished by offshore draggers, multimillion-dollar steel vessels about 160 feet (48 metres) in length, that tow a huge bag of net along the bottom that scoops up everything in its path. The northern cod stock normally produces almost half of Atlantic Canada's cod catch and a quarter of all the region's groundfish landings.

The same cod migrate to the shores of Newfoundland during the summer, supporting a once-abundant inshore fishery — but during the 1980s very few have been taken by the inshore fishermen's traps, nets and hooks. Until 1958, the inshore catch was never much less than 150,000 tonnes. By 1974, however, relentless overfishing by more than 20 nations had decimated fish stocks on both the Labrador banks and the Grand Banks, and the inshore catch had fallen to 35,000 tonnes. In 1977, Canada responded with a 370-kilometre (200-nautical-mile) exclusive fishing zone and a ban on foreign draggers. The inshore catch increased, peaking at 115,000 tonnes in 1982, but by 1986 it was down to 68,000 tonnes. Inshore fishermen were also complaining that the fish they did catch were unusually small.

The fishery remains "the last major industrial enterprise in the world in which raw material is still hunted instead of raised, grown, or produced,"

Stephen Kimber notes in *Net Profits*, his recent history of National Sea Products. The industry is tied to natural cycles as well as to economic and diet trends, but the situation in 1986 was not like earlier cycles: inshore nets were empty, while the offshore draggers were finding plenty of fish.

The roots of this paradox go back to 1977. With foreign draggers banned, Canadian fish companies embarked on an orgy of debt-financed expansion. In the early 1980s, however, interest rates and fuel prices soared, and most of the large companies found themselves insolvent. The federal government assembled Newfoundland's bankrupt firms into a supercompany called Fishery Products International. In Nova Scotia, new private investment and a massive infusion of government funds reshaped the venerable National Sea Products into a second supercompany.

Meanwhile, a federal task force headed by Dr. (now Senator) Michael Kirby devised a new quota system called "enterprise allocation." This gave the supercompanies fixed amounts of fish to catch and broke the cycle of glut and shortage that naturally resulted from unfettered competition for fish. Knowing how much they could take, the companies paced their fishing to maximize their profits, catching precisely what the market demanded and commanding premium prices.

Under enterprise allocation, National Sea went from losses of $17 million and $18 million in 1984 and 1985 to record profits of $10 million, $36 million and $25 million in the three following years. FPI did even better, earning $47 million in 1986 and $58 million in 1987, at which point its management privatized the company by buying back the government's shares.

The success of enterprise allocation suggested that science and sound management had tamed the fishery's cyclical nature, and the 1987 bonanza seemed to confirm those hopes. But 1987 was an exceptional year. Low world oil prices held the draggers' operating costs down, while a depressed Canadian dollar made Canadian fish cheap in the United States market. Consumers in the United States had learned to appreciate fish as a tasty, low-fat main dish with a whole array of health benefits. They consumed 20 percent more fish in 1987 than they had in 1983, driving prices to record levels.

Even during the euphoria, however, FPI's annual report noted that its trawlers had worked much harder to catch the same amount of flounder, and that its allocation of northern cod had been cut by 10 million tonnes. FPI's record sales were achieved because "price increases more than offset a reduction in the volume of raw material available to our company." Translation: we caught less fish, but we got more money for the fish we caught.

The northern cod reduction was particularly ominous. The Kirby task force had predicted that, by 1989, the northern cod stock would support a total allowable catch of a colossal 400,000 tonnes. To allow the stock to grow both in numbers and in total weight, the Department of Fisheries and Oceans set the catch not at the maximum sustainable yield, but at about 90 percent of that yield.

Yet even with this conservation measure, the allowable catch for 1988 was only 266,000 tonnes. The stock had grown much slower than expected.

Newfoundland's inshore fishery continued to fail. To the inshore fishermen, the reason was clear: fish were scarce, and offshore draggers were catching them before they came closer to shore.

Maybe not, said the scientists. Maybe the fish were staying offshore. Maybe fewer crews were actually fishing. Maybe it was the water temperature, or food cycles, or...

No, said the fishermen. There are not enough fish. And they were right.

By 1986, fisheries scientists were questioning their own statistics. With millions of dollars and thousands of jobs at stake, they checked their figures cautiously. In late 1988, they announced that their estimates of the northern cod population had been massively wrong. Their new figures warranted a total allowable catch for 1989 of only 125,000 tonnes.

"That was the greatest shock I ever had in the fishery," says Stephen Greene of Clearwater Fine Foods, the aggressive young company that grew up during the boom. "It was like someone telling you it's night when you're sure it's day."

An alarmed federal government asked Memorial University's Leslie Harris to chair a panel to review the scientific evidence. The panel's interim report, in May 1989, confirmed the bleak outlook. The northern cod stock, Harris explains, "has grown since 1977, but not nearly at the rate we thought, maybe not even 40 percent or 50 percent as much as we expected."

How could the scientists have been so wrong?

First, they do not have enough information. They do a certain amount of sampling themselves, but that sampling, Harris notes, is unreliable because it takes place at the same time every year. "If you assume the same conditions exist from year to year, then you're likely to be wrong. Fish don't follow our calendar."

To illustrate a more general sampling problem, Harris quotes Dr. Dayton L. Alverson of the University of Washington. Suppose you want to estimate the number of cattle on a ranch, says Alverson, but you can only do it in the dark, by towing a big bag across the ranch from a helicopter. In the morning you will see how many cows are in the bag, and you will estimate the total number of cattle from that. How likely is it that you will be right?

One standard way of measuring fish stocks is to calculate the catch-per-unit-of-effort. If a dragger tows its net for an hour, how many fish does it catch? If the number is steady from one year to the next and the next, then surely the

Is the foreign fleet just a scapegoat?

INTERNATIONAL law requires Canada to make any "surplus stocks" within our 370-kilometre (200-nautical-mile) zone available to other nations, with Canadian observers aboard to monitor the catch. These licensed foreign fishermen often pursue species whose very names are unfamiliar to most Canadians. Greenland halibut, for instance, occur far north of Newfoundland, while concentrations of silver hake are found at the edge of the continental shelf east of Nova Scotia.

Canada has used access to those surplus stocks as a reward for fleets that fish responsibly outside Canadian fisheries jurisdiction — on the Flemish Cap, an isolated seamount 1,000 kilometres east of Newfoundland, and on the "nose" and "tail" of the Grand Banks. With some nations — notably the Soviet Union — these bilateral agreements have worked well.

In theory, the outer fisheries are regulated by the 12-member Northwest Atlantic Fisheries Organization, but compliance with its regulations is voluntary, and several countries that fish the area are not members — including Mexico, South Korea, Panama and the United States. These nations operate, according to Canada's Standing Senate Committee on Fisheries, "with little or no regard for conservation."

In 1986, the European Community ignored its Northwest Atlantic Fisheries Organization quota of 23,260 tonnes and unilaterally set a quota of 102,460 tonnes. Its vessels, chiefly Portuguese and Spanish, harvested 172,183 tonnes. In 1988, the organization set the EC quota at 19,010 tonnes; the EC responded with its own quota of 163,100 tonnes, but the ravaged fishery produced only 66,395 tonnes.

The fish these countries pillage are often "transboundary" stocks, which move back and forth across the 370-kilometre limit. Some observers believe overfishing beyond Canada's limit may actually drain Canadian stock across the line.

In addition, France claims an overlapping 370-kilometre zone off the islands of Saint-Pierre and Miquelon, a dispute that has been referred to international arbitration. French fishermen "may have exceeded by four times" their quota in the waters south of Newfoundland, according to evidence heard by the Senate committee on fisheries.

Still, says Cabot Martin of the Newfoundland Inshore Fishermen's Association, foreign overfishing accounts for "no more than 20 percent of the problem." Foreigners are an easy target for Canadian politicians — but Canadians are mainly responsible for the crisis, and Canada will have to solve it.

S. D. C.

---

population is holding steady.

But the draggers — the measuring stick — have changed. Today's electronically equipped draggers are powerful fishing systems that can locate a school of fish, identify the species, lock onto it, and steer the nets into the thick of the concentration. When the holds are full, the crew can pinpoint the spot with satellite navigation systems and call sister ships to keep fishing while they nip ashore to unload.

Scientists knew the dragger fleet was evolving, but the changes came quickly — and catch-per-unit-of-effort, the standard measure, was not to be changed casually.

"Modern electronics tell the fishermen exactly where the fish are," says Dr. Harris, "and they fish on concentrations of spawning fish. If the fish are concentrated, then the catch will be as high as ever — and it will remain high until the last fish is caught.

"The state of our ignorance is appalling. We know almost nothing of value with respect to the behaviour of fish. We don't even know if there's one northern cod stock, or many, or how they might be distinguished. We don't know anything about migration patterns or their causes, or feeding habits, or relationships in the food chain. I could go on listing what we don't know."

However, in spite of the lack of knowledge, decisions have to be made. In February 1989, the Department of Fisheries and Oceans cut the 1989 quota for northern cod to 235,000 tonnes from 266,000 tonnes in 1988. Even that small a cut will cost the two supercompanies as much as $40 million. The 1990 total allowable catch has been set at 197,000 tonnes.

"That's too high," says Cabot Martin, a St. John's lawyer who represents the Newfoundland Inshore Fishermen's Association. "The ecological approach has to come first — and we'll somehow have to find the political will to take the pain."

Far to the south, on the Scotian Shelf, a different fishery faces the same sombre outlook: too much fishing power, too few fish. Here, too, inaccurate estimates of the fish stocks have produced unjustifiably high allowable catches, while individual fishermen and small processors have built boats and plants capable of handling four times more fish than is actually available.

On the Scotian Shelf, the focus falls not on a few score offshore draggers, but on 2,700 smaller boats that land their catches in dozens of villages tucked into crannies along the ragged, rocky coast. This armada is nearly impossible to police, and fishermen readily admit that cheating is easy and profitable. Nothing prevents a fisherman from delivering his fish to the local processor without reporting it — or, indeed, from trucking it directly to the Boston market. The Halifax *Chronicle-Herald* reports that up to 50 percent more fish may be taken than are reported; last year the Standing Senate Committee on Fisheries heard that the value of such illegal fish may have reached $100 million in 1985.

"There is no shame to getting caught and paying a $400 fine," a fisherman told the Senate Committee. "It is almost a badge of honour. I am sorry to have to be blunt with you, sir."

A Nova Scotia fishing executive is far less charitable. "We have to start seizing boats, withdrawing licenses and jailing poachers," he says, speaking off the record. "These bastards are criminals, and we should treat them that way. On the Scotian Shelf, in my opinion, we've got one year to turn it around. After that, we've lost the fishery."

But the fishermen themselves are caught in a vicious circle. Years ago, small vessels fished right along the coast, near their home ports. Today the nearby grounds are fished out, and inshore boats fish at Sable Island and Georges Bank, up to 160 kilometres at sea. That means bigger boats and more-advanced equipment, financed by hefty loans, with heavy monthly payments. Fishermen *must* pursue their catches relentlessly, with steadily growing technological sophistication. The long-term result is a devastated fishery, but the short-term result was sheer gravy; one Yarmouth fisherman told his lawyer that he did not know his 1988 income — but he paid over $65,000 in income taxes.

Greed and desperation are powerful motivators. The fisheries department tried to regulate the fishing effort indirectly by restricting the length of inshore boats to under 45 feet (14 metres). The complicated structures of wooden-hulled boats once ensured that a length restriction would automatically limit a boat's capacity. But the flexibility and strength of fibreglass has changed all that. A 35-foot (11-metre) wooden boat would have a capacity of between seven and nine gross tonnes, according to Coast Guard marine surveyor Harry Rex, but a fibreglass boat that length today would have a capacity of between 28 and 36 tonnes. Such boats are like fat bathtubs: clumsy, sluggish and unstable in a rough sea, but capable of carrying huge loads of fish.

And — oh, Canada — federal attempts at restraint are undercut by provincial policies aimed at expansion, like Nova Scotia's provincial subsidies for new fishing vessels. The provinces also license the processing plants, whose numbers increased from about 500 in 1977 to nearly 900 in 1988. Employment rose from the equivalent of about 25,000 full-time jobs to about 33,000 in the same period. Even after the plant closures of the past winter, Atlantic Canada's fish plants still employ several thousand more people than they did a decade earlier.

Fishing techniques also changed. Inshore boats traditionally used stationary nets or longlines with baited hooks at regular intervals. But 400 of the new inshore boats are draggers, costing up to $750,000 each. A decade ago, small inshore draggers could catch 23 tonnes per season; today's equivalent can catch 164 tonnes. A fleet of eight draggers owned by Yarmouth fisherman Lawrence Corkum and his sons is capable of catching 25 percent of the inshore quota for all of southwestern Nova Scotia.

All this fishing power has punished the stocks severely. The Department of Fisheries and Oceans reduced Scotia-Fundy's total groundfish quota from 279,000 tonnes in 1982 to just 168,000

Is aquaculture an alternative?

IF WILD STOCKS of fish are in decline, can aquaculture take up the slack?

Fish farming has certainly made great strides. This year, world production of farmed shrimp — largely from China — is predicted to equal the catch of wild shrimp. New Brunswick produced 5,500 tonnes of farmed Atlantic salmon in 1989, worth $50 million; Norway, the world leader, raised 140,000 tonnes. (For comparison, fishermen in British Columbia landed 85,000 tonnes of wild salmon in 1989.)

About 95 percent of Maritime mussels are farmed, and about 25 percent of oysters. Trout and even cod are now raised commercially in Atlantic Canada, and scientists and businessmen are investigating scallops, clams, lobster and halibut.

Atlantic Canada faces some obstacles in aquaculture, including fish stock supplies, financing and a limited number of sites. But it also has significant advantages in terms of scientific and technical expertise, competitive production costs, and proximity to the major markets of Eastern Canada and the northeastern United States.

Compared to the existing fishery, however, aquaculture will remain a minor factor in seafood production for the foreseeable future.

S. D. C.

The people of Canso fight back

IN 1607, two men stood talking on a Nova Scotia beach. Marc Lescarbot was a traveller, a writer and a legal adviser to Samuel de Champlain. Captain Savalet was the skipper of an 80-ton vessel from St-Jean-de-Luz, France, and his crew of 16 had taken 100,000 codfish that summer. A good season, he told Lescarbot, among the best in his 42 years of fishing at Canceau.

On January 7, 1990, 3,000 people marched through the streets of Canso, population 1,300. They stopped before the National Sea Products processing plant. Mayor Ray White said Canso was a symbol for one-industry towns everywhere. Fishermen's union president Pat Fougere said, "We're going to fight those arrogant bastards from Ottawa to Halifax!"

Fougere was referring to the federal and provincial governments and National Sea itself, which had announced on December 12 that it would close the Canso plant on April 2, throwing 750 people out of work.

Lying at the tip of a jagged peninsula thrusting into the Atlantic Ocean, Canso is the most easterly point in mainland North America. It has no prospect — and no intention — of being anything but a fishing port. The Canso plant is the largest employer in Guysborough County, and its closure would hurt all of eastern Nova Scotia. The school board would lay off one-third of its 300 teachers within two years of the plant's closure because of families moving away. The Chamber of Commerce in Antigonish, 100 kilometres to the west, acknowledges that 20 percent of that town's business comes from Canso residents.

But, as Mayor White noted, the company and the governments had "taken on a tough opponent in the people of Canso." In just a few months, National Sea and Clearwater Fine Foods had closed fish plants in the Nova Scotia ports of Port Mouton and Lockeport, and in St. John's, Nfld. National Sea's North Sydney plant had been geared down. In January, Fishery Products International announced closures at Trepassey, Gaultois and Grand Bank in Newfoundland.

But Canso seized the headlines. Canso people are fighters; in 1970, Canso's fishermen led an epic 15-month struggle to unionize the offshore draggers. The battle nearly precipitated a general strike, and contributed to the fall of G. I. Smith's Conservative government.

After the National Sea announcement, Canso erupted in fury. Governments have poured millions of dollars into a succession of companies to operate the Canso plant and its draggers, and the people of Canso were outraged at the idea of National Sea removing its publicly funded draggers and its publicly allocated fish quotas.

The community hired a management consulting firm to propose alternatives, it organized marches and petitions, and it sent delegations to Halifax and Ottawa to lobby company executives and government officials. Cape Breton Highlands-Canso MP Francis LeBlanc demanded — and got — an emergency debate in the House of Commons. On January 22, the Senate committee on unemployment insurance came to Canso to learn first-hand about the double whammy of plant closure and tighter rules for UIC.

As Mayor White said, Canso put up "one hell of a battle." On February 2, the provincial government announced a deal with National Sea that would keep 300 workers employed half-time, processing fish trucked in from around the region.

But Canso leaders — who had not been consulted — were not impressed. Without offshore vessels, they said, the plan simply represented a slower death. On February 17, more than 4,000 people gathered in Canso, roaring their opinions at a protest rally featuring New Democratic Party leader Audrey McLaughlin, Canadian Labour Congress president Shirley Carr, and Canadian Auto Workers president Robert White.

The next day, eight municipal politicians from Canso invaded Ottawa for an unprecedented round of meetings with the Liberal and NDP caucuses, the Commons Committee on Fisheries and Forestry, and federal ministers Barbara MacDougall, Elmer MacKay and Tom Siddon. Among other things, the Canso delegation proposed "community quotas" — a system that would grant fish quotas to established communities, in addition to quotas granted to corporations. Siddon was cool to the idea, but he promised to study the Canso submission and respond within a couple of weeks.

Four days later, Siddon was replaced as fisheries minister by Bernard Valcourt, who immediately met with the people of Canso but made no commitments. Canso, meanwhile, was considering a range of initiatives, from the establishment of a federal penal institution in the town to the creation of a Canso-owned fishing company to apply for unallocated quotas.

The continuing battle for Canso under the political and media spotlight raises one obvious question: What plight awaits the other communities in Atlantic Canada where the fish plant is the major employer?

S. D. C.

tonnes this year. Department information officer Joe Gough says that Scotia-Fundy suffers from "growth overfishing of cod and pollock," a level of fishing that captures all the natural increase in a fish population, preventing any growth. Haddock catches, however, have plummeted, and the stock is in danger of outright collapse.

The Department of Fisheries and Oceans tends to be more optimistic than fishermen and industry executives — but even Gough remarks that on the Scotian Shelf, "the big crisis may be yet to come."

For many fisheries observers, draggers are the common denominator in all the various crises. Elnathan Smith manages a 50-year-old family fish business in Shag Harbour, N.S. He has spent his whole life in the fishery, and says flatly that "there are too many draggers, both offshore and inshore."

"When a drag tows through the bottom of the ocean," Smith wrote to then Fisheries Minister Tom Siddon in August 1989, "it collects anything and everything that is in its path, regardless of size or species. When the drag is dumped on deck, only the species directed for is kept, and in some cases only the large fish of that species are kept because of the higher prices paid. The rest is thrown back into the ocean." This bycatch, of course, is dead.

When draggers are used, the sea bottom "is scored as though with plough-shares, and rammed down as though with steam-rollers," says former Soviet dragger captain Vladil Lysenko, who fished extensively in Canadian waters. "Nothing is left alive for the fish to eat. What is more, this is where the fish breed, and when they lose their breeding grounds, the fish die out without leaving any progeny."

The baited hooks of longliners, by contrast, automatically regulate the size of the fish they catch and there is no bycatch.

Draggers perfectly symbolize an unsound and uncaring approach to the fishery — and indeed to the environment at large. Scouring the ocean floor, the dragger indiscriminately captures big fish and small ones, dogfish and monkfish and sculpins, spawning fish and juveniles, plants and lobsters and rubber boots and tin cans along with the occasional skull or thighbone of a drowned sailor. Like the tree harvester and the dynamite stick, the dragger applies raw force to a complex and sensitive ecological system. The fishery may well be a symbol for the environmental predicaments we face on many fronts.

Drastic reduction or even abolition of draggers is a feature of many plans for reform. Clearwater's Stephen Greene and a colleague have published a radical blueprint for the Scotian Shelf, including a ban on draggers except in rare circumstances where no other technique can be used — with flounder, for example, which do not easily take the hook. Greene's approach also includes punitive fines for overfishing, a ban on gillnets, a zero quota for haddock, and an industry-wide emphasis on longlining.

It is hard to imagine a recovery of the fishery without tough, rational regulations of this kind. For some stocks, though, it may already be too late.

Considering this point, Bob Lee, a Coast Guard supervisor in Dartmouth, N.S., shakes his head gloomily. He cites the once-rich California pilchard fishery — similarly plundered, similarly regulated, similarly poached.

"Today," he says, "there is not one single pilchard left on the coast of California. A free resource is always exploited to extinction, for the minimum possible benefit — and that is where the Atlantic fishery is headed."

Atlantic Canada's daunting task is to prove Lee wrong.

Silver Donald Cameron is a Nova Scotia writer who contributes regularly to Canada Geographic.

Sky patrols over the Grand Banks

High-tech eyes keep watch on foreign fleets off Newfoundland

By Michael Clugston

AS THE AIRPLANE sped rapidly closer to the unidentified vessel showing as a blip on the radar screen, the flight crew and passengers tensed in their seats. "Half a mile to the contact," called out radar operator Ken Ludlow. "I'll bring the ship up on the right side so Frank can see it," he said into his microphone. His words were heard inside six earphones, by six occupants of the speeding Beech Super King Air 200. Tearing through the dense cotton of a Grand Banks fog at 320 kilometres an hour, plane and occupants relied entirely on the seeing-eye skills of Ludlow's technology.

"Turn five degrees left," said Ludlow. "Five hundred feet to target, four hundred feet...and...turn NOW." The plane executed a precise, thrilling pirouette on its right wingtip, directly over the invisible ship. Six pairs of eyes strained to penetrate the fog that blanketed the wavetops 50 metres below. But the "target" just outside Canadian waters — probably a foreign deep-sea trawler — might as well have been kilometres away. "It's no good, boys," said Frank Snelgrove, the fisheries officer on board. "We won't see her today. Not in this thick stuff. Let's move on — we may find a hole in the fog somewhere else."

Three hundred and seventy kilometres out at sea, over the Grand Banks of Newfoundland, the twin-engine King Air pulled above the fog bank and roared into the sunlight at an altitude of about 400 metres. Pilot Bruce Jollymore set the plane on course for the next contact on the radar screen, and the federal government's fisheries patrol hoped, yet again, for a break in the fog. "At least they know we're in the area — no other plane would be here," said Snelgrove.

Fisheries officers are the game wardens of the Grand Banks of Newfoundland, and the surveillance patrols are conducted by the Department of Fisheries and Oceans and the Department of National Defence. Physically, the famous banks are simply a shallow undersea plateau once crowded with an unbelievable bounty of fish. But to fisheries regulators they are a complex patchwork of fishing zones and quota limits, gear and season restrictions — the arena for the struggle between the opposing impulses to fish and to conserve.

The fisheries patrols have a divided, frustrating role. Inside the Canadian 200-mile limit, Canadian and some foreign fleets are licensed to catch various species of fish, and the patrols try to prevent violations such as using illegal gear, fishing in the wrong areas, over-fishing of quotas, and mis-reporting catches. Their workload is eased by federal fisheries observers, who monitor the fishing on board all large off-shore vessels in Canadian waters. But outside Canada's zone, quotas are established by the 12-member Northwest Atlantic Fisheries Organization (NAFO). In these international waters, Canada's fisheries patrols are empowered to observe but not to enforce — NAFO has no enforcement powers.

Through an accident of geography, these banks are among the few places in the world where the continental shelf extends beyond a nation's 200-nautical-mile marine boundary. Two fish-rich corners of the shelf, the "Nose" and "Tail," lie in international waters beyond the reach of Canada's fisheries management programs. At any given time, a fleet of from 75 to 250 foreign vessels can be found clustered tightly over those two areas.

In the last five years, this armada has caught some 500,000 tonnes more than the quotas established by NAFO — a level of depredation widely blamed for much of Newfoundland's troubles in the current fisheries crisis that has left some 8,000 fisheries workers unemployed in Newfoundland and Nova Scotia. The worst offenders are from Spain and Portugal, two nations within the membership of the European Community (EC).

"For all intents and purposes, the EC vessels are fishing out there unrestrained and the only thing that limits their catch is whether or not there's fish available to catch," says Les Dean, Newfoundland's assistant deputy minister of fisheries. That is a fair assessment, reflected in the fact that the EC fleet has not been able to catch the inflated quotas it unilaterally sets for its own fleet.

These "straddling" stocks swim back and forth across the 200-mile line: overfishing outside the line reduces the catches inside. Foreign overfishing focused first on the Tail, from 1986 through 1990, and Canadian catches of key stocks were devastated: southern cod catches dropped 44 percent, flatfish 55 percent and flounder 67 percent. Once that fishery was depleted, the foreign fleet began to concentrate more on the Nose, targeting the northern cod stock. Northern cod live for most of their life cycle inside Canadian waters: only a small proportion of the stock swims into international waters each year to spawn. This is the most important fish stock in Canada's Atlantic fishery, supporting some 31,000 jobs and bringing an estimated $700 million to the Canadian economy in 1991. Although NAFO has declared a moratorium on fishing the stock outside the Canadian zone, foreigners took about 47,000 tonnes last year. Inside the zone, Canadian vessels were only able to find and catch two-thirds of their 185,000-tonne quota.

The Canadian surveillance teams can only arrest and detain these foreign ships when they slip illegally into Canadian waters, where fish are relatively more plentiful, to take a quick catch of fish before racing back across the line. Since 1985, Canada has laid more than 100 charges against foreign trawlers for boundary violations and other infractions in the Newfoundland region, trying them at the Provincial Court in St. John's. In reaction to increased foreign activity, sea and air patrols were increased in 1991, under the $584-million Atlantic Fisheries Adjustment Program, from two overflights a week to nine, and 800 at-sea inspections in 1991 compared to 340 the year before. The increased patrols made their presence felt: in 1991 there were only seven border incidents and 11 other violations.

"Most violations come at night, in fog," says Leo Strowbridge, coordinator of offshore surveillance for DFO. Surveillance is a game of cat and mouse played out by the Canadians with radar, cameras, computers, satellite data — and by the Europeans and others with quick sorties under cover of darkness and poor visibility. The yellow and green King Airs fly in fog, in snowstorms, and even at night with a 15 million candle-power searchlight sweeping the ships to read the identification numbers painted on the sides. The flights leave at irregular, unannounced times, keeping radio silence so poachers won't hear them coming.

A computer buff's dream come true, the three planes based in St. John's and Halifax (to patrol the Scotia-Fundy region) are flying data centres. A case of illegal fishing must be recorded so thoroughly that the evidence will stand up in court months later — the point of the whole exercise. That evidence includes photographs and computerized data stored in an airborne system that links navigation, radar and computer memory, called the computer controlled navigational system (CCNS).

"We had to achieve a few 'firsts' to make the CCNS work," says Robert Halliday, operations director at Atlantic Airways, the private firm that owns and operates the planes and contracts them to DFO. The firsts were to mount a PC desktop computer on board, shield it from the plane's vibrations, and link it with two navigational systems, Loran C and the satellite-based Global Positioning System. Then they added the most advanced radar available at the time — the system was put together in 1986 and 1987 — the Litton V-5, capable of spotting a small vessel 320 kilometres away. "There are so many antennas on the outside of the plane that it's a real challenge for the technicians to mount new ones so they won't interfere with each other," says Halliday.

"In the early 1980s, before we had this system, the courts would tolerate unlicensed foreigners 10 miles inside the line," says Strowbridge. "They didn't think our navigation was any more exact than that. But with this sophisticated technology, the courts have now accepted a tolerance of less than a kilometre."

It is one thing for the King Air's team to spot offenders, but the foreign captains can only be brought to court by fisheries officers going aboard and forcing the ships to head for St. John's.

The Spanish trawler *Amelia Meirama* was making a run for international waters, its bow deflecting bursts of icy spray as it plowed and rolled through the three- to four-metre swells. Its lights were off, since the vessel was fishing illegally just inside the Canadian 200-mile zone, beside its companion vessel *Julio Molina*. It was 4 a.m. on May 22, 1986 — a moment that Canadian fisheries officers Ben Rogers and John Taylor, in their late 20s, will never forget. They were desperately gripping the trawler's rail, their feet dangling in the air a metre or so above the icy waves, and they were weighted down with flashlights, lifejackets, survival suits — regulation gear for apprehending a foreign trawler and taking it into St. John's to face charges. A moment earlier Taylor and Rogers had lept to the rail from their small rubber dingy from the DFO patrol ship *Cape Roger,* following at a safe distance.

"It was scary for a minute to be hanging there, in the complete dark, 200 miles from land," recalls Rogers. "If you drop into the water from any height, you go under for a while. It's pretty easy for the guys in the following DFO vessel to lose sight of you." What's more, the trawler was pulling its net, which could have entangled a person in the water.

Taylor felt around in the dark with his boot and managed to get a purchase on a porthole. He reached back and grabbed Rogers by the scruff of his rubber survival suit and helped him up. When they confronted the astonished captain on the bridge, he refused to travel to St. John's: instead, he set course for his home port — in Spain. By the time the Canadian patrol ship *Leonard J. Cowley* overtook both trawlers with an armed RCMP boarding party, they were 300 kilometres from the Azores Islands — 1,100 kilometres from St. John's. The ship was forced to return to St. John's, and the captain was tried in Newfoundland Provincial Court. The owners were fined $150,000, $38,000 in fish was confiscated and the ships were detained for days. The penalty was roughly equivalent to the value of 50,000 tonnes of fish, and the average trawler takes about 1,500 tonnes on a voyage. Are such penalties deterrents? They may very well be — only two boundary line charges were laid in 1991.

"For those three days, the Spanish fed us and were quite cordial," says Rogers. "They said they knew we were just doing our job. As the captain went into court he said to me, 'Ben, you're my friend, but in court you're not my friend.' I said, 'Captain, I understand.'"

After that incident, the federal government armed its officers with 9mm Heckler and Kock machine guns, to be used only in boundary line violations. The new firepower has turned the boardings of foreign vessels into much more tense police actions.

The Newfoundland government estimates that in 1990 and 1991 alone, the worsening fishery has cost Newfoundland close to $150 million in lost exports: it blames the foreign fleet for 75 percent of the losses. Its proposed solution is to impose unilateral control, or custodial management, over the Nose and Tail.

"There is a framework under the Law of the Sea Convention in terms of custodial management on behalf of all nations," says Les Dean. "It means we would manage these fish stocks on behalf of the international community, sharing the benefits of that management."

The proposal — which would put the surveillance crews on an even hotter hotseat — is taken seriously in many quarters. "This is not a whimsical proposal," says Judith Swan, a fisheries lawyer and executive director of the Oceans Institute of Canada, a nonprofit, independent organization based in Halifax. "It's not as if Canada decided unilaterally that the Europeans have been fishing too much. Canada has made representations in every available forum for at least five or six years, trying to reduce the overfishing. In order to provide for the continued existence for the stock as a whole, it is justifiable to say we're going to manage it out there — even on a temporary basis."

The federal government continues to favour a diplomatic solution, but Fisheries Minister John Crosbie says if that does not work, "...other measures, including unilateral measures, will have to be contemplated," Federal diplomacy will be directed through a special session of NAFO in May, and in talks at the United Nations Conference on Environment and Development in June. But many observers see the federal government as part of the problem. "A major part of the issue has always been our own overfishing as Canadians," says Cabot Martin, spokesman for the Newfoundland Inshore Fishermen's Association. "We can't throw rocks at the foreigners until we clean up our own act."

Martin and many others argue that high Canadian quotas have been a major cause of the current crisis. In an authoritative 1989 study, Dr. Leslie Harris, president of Memorial University, warned that the northern cod stock could be eradicated unless the allowable catch was cut sharply to 100,000 tonnes — less than half the 1988 limit of 266,000 tonnes. But the federal government instead put its priority on jobs, setting the 1991 catch at 190,000 tonnes. Although the northern cod quota for the 1992 season was dropped 35 percent to 120,000 tonnes, that is scarcely less than Canadians were able to catch in 1991. "It's really to say we're going to allow fishermen to continue to catch all that they're capable of catching," Harris told a radio interviewer last winter. Still, the

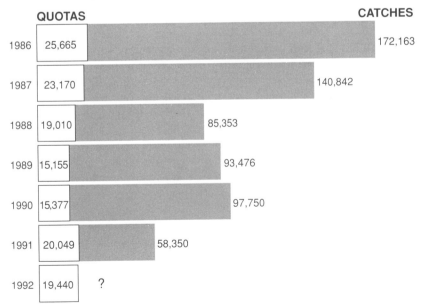

European Community's NAFO quotas vs. actual catches
(tonnes of northern cod and other groundfish)

Year	QUOTAS	CATCHES
1986	25,665	172,163
1987	23,170	140,842
1988	19,010	85,353
1989	15,155	93,476
1990	15,377	97,750
1991	20,049	58,350
1992	19,440	?

fact remains that Canadian conservation measures will be defeated by foreign overfishing on the same stocks.

Harris is an impartial observer of the highly politicized and fractious East Coast fishery. He favours Newfoundland's proposed custodial management of the Nose and Tail. But he also says it is unclear whether there are enough northern cod left to regenerate the stock. "My reaction is that the quota reduction is too little, too late," he says. The number of cod mature enough to reproduce has plummeted drastically, by as much as 50 percent in 1991, according to a report last winter by the Canadian Atlantic Fisheries Scientific Advisory Committee.

The fisheries patrol plane swings home on a route over the jagged, rocky pan of the Avalon Peninsula. To the south and west, the processing plants are closed in the single-industry towns of Grand Bank, Gaultois, Trepassey, Arnold's Cove and Fortune. To the north, in Catalina, 1,000 fishery workers were laid off last winter for lack of northern cod. The plane flies low over St. John's harbour, where the National Sea Products Ltd. plant laid off 170 workers last March.

Reckless fishing has destroyed many once-prosperous industries: herring in the North Sea; anchovies off Chile; and haddock, halibut and other species on the Grand Banks before Canada extended its boundary to 200 nautical miles in 1977. Is Canada's northern cod next in line? The answer will hinge in part on the offshore surveillance teams, since flying the flag seems to make a difference. "Letting the foreigners know we're out there does affect the way many of these skippers operate," says fisheries officer Snelgrove as the airplane touches down on the tarmac. "I went aboard one Spanish trawler and the skipper said to me, 'If I don't see your plane for three or four days, I'll go inside the line maybe four, five kilometres. What would you do if you caught me?'

"I told him we'd slap the cuffs on him," says Snelgrove. "He threw his head back and laughed. About three weeks later we caught him doing just that. He's gone through the preliminary hearing stage, and is now committed to trial in St. John's." Even for the harried "game wardens" of the Grand Banks, life has its small satisfactions. ◆

Michael Clugston is senior editor of Canadian Geographic.

JOHN C. CROSBIE
Minister of Fisheries and Oceans

Can we stop the foreign plunder of Grand Banks fish resources?

For the second time in recent decades, one of the great food resources of the world faces a grave and growing crisis. Over-harvesting by foreign vessels is drastically depleting the stock of fish that live over the Grand Banks off Newfoundland. Foreign vessels fishing just beyond Canada's 200-mile exclusive economic zone are taking annual catches many times greater than internationally agreed upon quotas. One effect is a sharp reduction in the resources on which Canada's fisheries industry depends. International law permits Canada to take unilateral action in the high seas to protect its fisheries if international negotiations fail. Canada continues to seek international action to stop the over-fishing, but time is running out for the fisheries resources and those who depend on them. Tangible progress must be made in 1992, or Canada will have to seriously consider unilateral action. Address to the Royal Institute of International Affairs, conference on international boundaries, London, January 10.

I am grateful to the Royal Institute of International Affairs for the opportunity to address this distinguished and learned gathering on the topic of "Boundaries and Fisheries."

I will focus very largely on the situation in the Northwest Atlantic off Canada's coast. The problem I will speak about is overharvesting on the high seas of fish stocks that straddle the 200-mile limit. This occurs elsewhere, such as in the Bering Sea between the United States and Russia. But it is in the Northwest Atlantic that the problem has become most critical in terms of impacts on coastal communities, and consequently where measures to end over-fishing outside 200 miles are most urgently needed.

I approach my topic today from three perspectives: first, ecology and sustainable development; second, the practice of states fishing in the Northwest Atlantic; and, third, rights and obligations under international law. It is only by bringing together all these aspects of the problem that we can understand and seek to solve it.

As the 1987 World Commission on Environment and Development, more commonly known as the Bruntland commission, stated:

"Without agreed, equitable and enforceable rules governing the rights and duties of states in respect of the global commons, the pressure of demands on finite resources will destroy their ecological integrity over time...

"Only the high seas outside of national jurisdiction are truly 'commons,' but fish species, pollution and other effects of economic development do not respect these legal boundaries."

Put in other words, should not freedom of the high seas mean something different than a right to over-fish and licence to pollute?

In 1990, Dr. Lee Alverson, a leading authority on world fisheries, in an address on *Fisheries Management: Transition to the 21st Century,* described the pattern of global over-exploitation of fisheries that has proceeded apace since World War II. He identified as key problems over-capitalization of fleets, competition for trans-boundary stocks, and what he called the jurisdictional void for management of resources beyond 200 miles.

These problems become more acute as coastal states puts in place more effective management and control measures for stocks within 200 miles. Distant water fleets dis-

Courtesy of the Ministry of Fisheries and Oceans

placed by such measures fish wherever they can, increasingly outside 200 miles. This is part of what is happening off Canada's Atlantic Coast.

The fishing area I am speaking about is the Grand Banks of Newfoundland. This is a large area of relatively shallow water — generally less than 100 metres — extending east and south from the coast of Newfoundland. One part of the Grand Banks extends out to 250 nautical miles; this is commonly known as the Nose of the Bank. Another part extends out to 260 nautical miles; this is known as the Tail of the Bank. The Grand Banks total over 100,000 square nautical miles, of which the Nose and Tail total only 12,000 square nautical miles.

Although migratory patterns differ from species to species, one element is common to the cod, flounder and redfish stocks that straddle the 200-mile limit: while they live primarily inside 200 miles, many fish migrate for part of the year outside 200 miles to deeper water on the edge of the shelf. There they are subject to intensive over-harvesting that has led to stock depletion.

This is not supposed to occur. Rather these straddling stocks are supposed to be harvested in accordance with conservation management decisions made by NAFO, the Northwest Atlantic Fisheries Organization. Canada, the EC, Japan, the USSR and Norway, among others, signed the NAFO Convention to create the organization in 1978. NAFO was established to scientifically assess stocks, decide on a total allowable catch, allocate quotas to NAFO contracting parties based on established sharing arrangements, and set inspection and control measures for the area outside 200 miles. However, it remains for NAFO contracting parties, exercising flag jurisdiction over their fleets, to give effect to NAFO decisions.

Before 1986, the EC [European Community] adhered to all NAFO decisions. In 1986, Spain and Portugal joined the Community. They were given limited access to EC waters, but promised other fishing opportunities. Beginning in 1986, the EC has objected to various NAFO quotas and set higher quotas for itself unilaterally, quotas that have been allocated largely to Spain and Portugal.

The EC was able to do this by misusing a provision in the NAFO Convention, whereby contracting parties can file an objection to decisions they do not wish to adhere to. Instead of being used by a state to avoid being discriminated against, as the provision was intended to be used, the EC has used it as an ongoing means to allocate itself a greater share of fish stocks. Thus, instead of being used to avoid discrimination, the objection procedure has been misused by the EC to gain a preference.

As well, the EC has failed to exercise effective control over its Iberian fleets, which have fished virtually at will outside 200 miles. As a result, catches by EC fleets have greatly exceeded both NAFO quotas and unilateral quotas the EC has set for itself.

This has created a crisis in fisheries in the Northwest Atlantic. From 1986 to 1990, EC fleets reported catches of cod, flounder and redfish totalling more than 530,000 tonnes for the five years. This was more than five times the EC's NAFO quotas that totalled for 1986 to 1990 just under 100,000 tonnes. For some flounderstocks, the EC's reported catches exceeded its NAFO quotas by 18 times, much of this being juvenile fish harvested in nursery areas.

These are reported catches. They do not include un-reported catches or catches mis-reported as un-regulated species, like skate. Nor do these figures include catches by non-NAFO fleets, which include Korean and other vessels, but are largely comprised of Spanish and Portuguese vessels operating under flags of convenience. From 1986 to 1990, non-NAFO vessels, operating without any NAFO quotas, caught over the five years more than 165,000 tonnes of cod, flounder and redfish.

While the figures for EC catches in 1991 are not yet complete, they will not be markedly different from recent years, notwithstanding the EC's acceptance of most NAFO quotas for 1991. In the face of this continued overfishing, fish stocks have been seriously depleted and, as a result, many quotas have been cut in half. From 1986 to 1990, Canada's quotas for regulated species over-fished by EC and non-NAFO fleets were reduced by a total over the five years of 300,000 tonnes. Quotas for other NAFO contracting parties like Japan, Norway and the former USSR, have plunged as well.

The pattern is now well established. Canada and other NAFO contracting parties, except the EC, limit catches by their fleets to conserve fish stocks, but EC and non-NAFO fleets take all they can outside 200 miles. Canada and other NAFO contracting parties then further limit their catches in line with quotas reduced to reflect stock depletion. But EC and non-NAFO fleets continue to take all they can, often just outside the 200-mile limit.

The NAFO Scientific Council, comprised of scientists from various NAFO contracting parties, concluded in its 1991 report that under-reporting and mis-reporting of catches has become so prevalent that it is no longer possible to do scientific assessments of NAFO-managed fish stocks.

What is known is that depletion of fish stocks is continuing under the pressure of heavy overfishing. We are now flying blind, certain only that the crisis is becoming more grave and that an effective response is urgently needed.

While Canada has had to reduce its offshore fleet by about one-third since 1986 — and EC fleets are being reduced in Community waters in an effort to end over-fishing there — the number of EC and non-NAFO vessels fishing in the Northwest Atlantic has increased? Why? Because the area just outside 200 miles is one of the few convenient fishing areas where distant water fleets can continue, in effect, to fish at will.

For example, a new Spanish fleet appeared in the Northwest Atlantic in 1990, comprised of vessels driven out of Namibia's 200-mile zone because of overfishing there. The catching effort of this fleet was simply transferred from inside 200 miles off the West Coast of Africa to outside 200 miles in the Northwest Atlantic. Why? Because off the West Coast of Africa the 200-mile limit falls outside the migratory range of major fish stocks, while in the Northwest Atlantic the migratory range extends to the edge of the continental shelf, which lies just outside the 200-mile limit.

Canada has made repeated efforts since 1986 to persuade the EC and its member states to end their over-fishing in the Northwest Atlantic. While there have been some efforts by Brussels to control EC fleets, there has been no appreciable reduction in EC catches of regulated species in the Northwest Atlantic. And time is running out for the stocks being over-fished.

As well, given the increasing resort to flags of convenience, the pressure on fish stocks has grown worse. The use of flags of convenience could render futile conservation efforts by all NAFO contracting parties, including the EC. It matters not a whit whether over-fishing is carried out under the flag of Spain or Panama or Korea if the stocks continue to be depleted.

Efforts to manage international fisheries in the Northwest Atlantic to avoid resource depletion date from the time when the introduction of modern harvesting technology made depletion a real possibility. In 1949, Canada and other concerned states signed the International Convention on Northwest Atlantic Fisheries (ICNAF). ICNAF operated until 1978 when it was replaced by NAFO.

> *"Tangible progress must be made in 1992 or options that no state has wanted to consider will have to be seriously considered."*

ICNAF was singularly unsuccessful in managing the fishery, as high seas resources — then outside three and later 12 miles — were over-harvested on a scale that dwarfs the current problem. The resulting resource depletions were dramatic, with reductions of as much as 80% in major fish stocks. The ICNAF period demonstrated by harsh experience what was needed to conserve and manage resources, lessons that we have sought to apply in NAFO.

There is a grim sense of deja vue today, as fish stocks that were rebuilding toward healthy abundance in the early NAFO years — 1978 to 1986 — are now declining for the second time because of high seas over-fishing. We have suffered the harsh consequence of over-fishing once before and we do not wish to suffer them again.

High seas over-harvesting worldwide gave rise to repeated efforts to develop international law to deal with the problem. It is useful to recall that the 1958 Law of the Sea Convention provides in Article 7 that,

"any coastal state may, with a view to the maintenance of the productivity of the living resource of the sea, adopt unilateral measures of conservation appropriate to any stock of fish or other marine resources in any area of the high seas adjacent to its territorial sea, provided that negotiations to that effect with the other states concerned have not led to an agreement within six months."

These provisions were not used in the Northwest Atlantic, but rather Canada directed its efforts through multilateral means in ICNAF and NAFO.

Let me turn now to legal rights and obligations under the 1980s Law of the Sea Convention. Article 87 provides for the freedom of the high seas, including freedom of fishing. However, this is subject to obligations imposed by the Convention to co-operate in the conservation and management of living resources. Article 63 provides that for straddling stocks, coastal states and distant-water fishing states shall seek to agree on conservation measure applicable beyond the EEZ (exclusive economic zone) either directly or through appropriate regional organizations. For the Northwest Atlantic, NAFO is the relevant regional organization.

But what happens if the decisions of the relevant regional organization are not adhered to? And what happens if efforts made in good faith by the coastal state bilaterally with distant-water states do not end or even reduce the over-fishing of straddling stocks outside 200 miles? Does the coastal state have a right to the fisheries resources within its 200-mile limit, but no remedy when that right is negated by over-fishing outside 200 miles?

As Alan Beesley, the chairman of the Drafting Committee for the 1982 convention, wrote in 1988 in *The Future of International Oceans Management:*

"[in considering] the conservation of the living resources of the oceans and the preservation of the marine environment... we must not only take into account the fisheries and environmental provision of the [Law of the Sea] Convention, but the extent to which they are being implemented or ignored... When it comes to the high seas beyond national jurisdiction, international action is essential."

In 1990, Canada launched an international legal initiative to implement and render effective the provisions of the Law of the Sea Convention concerning straddling stocks by clarifying their operation. We are seeking to achieve the purpose behind those provisions, that is effective measures for conservation and management of resources both inside and outside 200 miles.

Canada has adopted a pragmatic, multi-track approach. We are pressing for agreement with NAFO for additional surveillance and control measures, like a joint observer scheme and reciprocal enforcement arrangements. We will continue our efforts bilaterally with the EC and with flag states for non-NAFO fleets. But increasingly it seems that developments in international law are needed.

Canada is calling for the adoption of principles and measures, within the framework of the Law of the Sea Convention, toward ending over-fishing and straddling stocks outside 200 miles. Here are some examples from proposals tabled by a group of concerned coastal states for the U.N. Conference on Environment and Development.

1) High seas fishing must be carried out on a basis of sustainable, ecologically sound practices, effectively monitored and enforced, in order to ensure conservation and promote optimum utilization of living resources.

2) High seas fishing must not have an adverse impact on the resources within the jurisdiction of coastal states.

3) States must effectively monitor and control fishing activities of their nationals, vessels and crews on the high seas to ensure the conservation of resources, compliance with applicable conservation and management rules, complete and accurate reporting of catches and effort, and avoidance of incidental catches.

4) In areas of the high seas where a management regime has been agreed within the framework of a competent international organization — states must ensure that high seas fishing may be undertaken only in accordance with the conservation and management rules adopted under that organization.

5) With respect to a stock occurring both within the exclusive economic zone of a coastal state and in an area of the high seas adjacent to it, the management regime applied to the stock must provide for consistency of the measures applied on the high seas with those applied by the coastal state within its exclusive economic zone.

None of this is particularly radical, yet it is being resisted by a small number of states operating distant-water fleets because they say it will derogate from freedom to fish on the high seas. The alternative to Canada's approach is unilateral action outside 200 miles. Such actions and their possible consequences are something that no state has wanted to consider.

Let me close on a personal note. For three years, I have played a leading role in Canada's efforts toward ending over-fishing of straddling stocks. Over that time, much has been accomplished to increase international awareness of the problem, but little progress has been made toward achieving new measures to avoid further depletion of straddling stocks.

And every year more vessels fish outside 200 miles, their catches remain at levels that cannot be sustained, and straddling stocks are further depleted. Time is running out for the fisheries resources and those who depend on them. Tangible progress must be made in 1992 or options that no state has wanted to consider will have to be seriously considered.

Canada will continue to take a pragmatic approach and will pursue all available avenues toward a solution. We are calling for international support for principles and measures that will solve the problem within the frame work of existing international law. In doing so, we are seeking what the Brundtland Commission called for, "agreed, equitable and enforceable rules governing the rights and duties of states."

Haddock getting healthier

Fisheries experts in Nova Scotia are finding that, contrary to the general trend in Atlantic Canada, the haddock population in the Georges Bank area is healthier now than it was ten years ago. Not only are fish more plentiful, but their average size has also been increasing.

Historically, fishing rights on Georges Bank, an area of ocean between Cape Cod and Nova Scotia (at the southwest end of the Scotian Shelf) have been hotly debated between Canada and the United States. One effect of the dispute was depletion of the fish stocks, whose number and distribution decreased considerably.

In 1985, the International Court of Justice established a geographic boundary to settle the fisheries question. Canada was given the rights to control a larger portion of the haddock's territory than the United States. American boats, which didn't have quotas imposed on them (they still don't), could no longer fish in waters east of the new fishing zone.

Canadian fishers saw their catch go from under 1500 tonnes in 1984 (before the boundary was set) to over 3500 tonnes in 1991. During the same period, the catch for American fishers went from the same level as Canada's, to less than 1000 tonnes in each of the last three years. The US no longer fishes directly for haddock.

Experts say that the marked difference between the catches for the two countries is mainly due to lack of controls on the American effort (leading to stock exhaustion), and better management by Canada. "There is suitable habitat on the US side...there isn't any ecological reason why the fish distribution would have such a marked difference across the boundary," says Stratis Gavaris, fisheries biologist with the Bedford Institute of Oceanography.

In 1985, the average age of fish caught by Canadians was two years. Consistent with the premise that fish stocks are now healthier, the average age has risen to between three and five years. "When you are dependent on several instead of one year-class of fish, catches are more stable and that should increase stability in the industry," Gavaris says.

Haddock is fished in the summertime when the trawler and longliner fleet can take advantage of the milder weather to venture further offshore. The reliability of the season is important to many fishers who need the high prices commanded by haddock to help increase their financial stability. ❏

Courtesy of Alternativess

READING 6

1. This reading demonstrates that exploited fisheries exist all over the world for the same reasons: greed and the rapacious harvesting ability of modern fishing ships. It's also a matter of property rights. What does the author mean by this?

FISHING FOR TROUBLE

BY HAL KANE

Back in the 1950s, marine biologists at the U.N. Food and Agriculture Organization (FAO) issued a prophetic warning: there is a limit to how much fish the oceans can yield. At the time, about 30 million tons of fish were being caught each year, and the figure was heading upward. The biologists calculated the oceans' sustainable harvest at about 100 million tons per year. Anything above that level, they suggested, would invite later declines, since fish hauls would exceed the rate at which fish reproduce.

Three years ago, that hypothetical ceiling was reached, and now it appears much of it evidently stemmed from past fishing that outpaced many species' ability to regenerate.

If the FAO biologists were right, annual catches may fluctuate around the 100-million-ton level from now on, despite massive future increases in human population and demand for food.

The dilemma of over-fishing is that the harder fishermen work for their catch, the fewer fish there are to take, and the more threatened their livelihoods become. After buying new "high-tech" boats throughout the 1980s that could catch more than ever, New England fishermen have seen their catches diminished drastically due to overfishing. Today, they must work long days and pursue "trash" fish like hake, whiting, spiny dogfish, and skate, which were previously spurned. Once-plentiful cod, flounder, haddock, and redfish are so rare that they're not worth pursuing even though they fetch higher prices.

The powerful new ships off Alaska's coast have the capacity to take a year's quota for the region in less than six months, and must sit idle the rest of the time. A year's quota of Pacific halibut is now taken in two frantic 24-hour periods, says *U.S. News and World Report*. That spells bad news for fishermen who need the work and now must pursue new careers and new ways of life.

Farther to the north, in the once-rich waters off the coast of Newfoundland, the decline in the cod harvest is so serious that the Canadian government recently suspended all cod-fishing — a crushing blow to the regional economy. And across the Atlantic, Icelanders have been told that the island's cod stock could collapse unless the annual catch quota is cut back by 40 percent in 1993. Cod makes up about a third of Iceland's exports, and the choice between losing jobs to restrictive quotas and losing jobs to a collapse of the cod fishery will be an unhappy one. But ultimately, fewer jobs will be lost if quotas protect the fish and let them reproduce than if a lack of quotas brings declining stocks and the ruin of a major industry.

Off the coasts of every continent, fish depletions have become apparent — partly because commercial fishers are sailing farther afield as prospects in their home waters become limited. European fleets have turned toward southern oceans, and are blamed by some African nations for reducing catches by as much as two-thirds in recent years.

Namibia, for example, complained during the 1980s that the invasion of European fishers was robbing it of both food and food-processing jobs. To preserve its fishing industry, the newly independent country banned all foreign ships from its waters, established fish-processing industries on its own territory, and stiffened quotas to allow pilchard, anchovy, hake, and other commercially valuable fish to regenerate.

At the center of efforts to maintain the world's fish supplies is resolution of the issue of property

Courtesy of World Watch

rights. Since fishermen can't own a portion of the oceans, they often have no understanding of their stake in preserving what swims beneath the surface. Rather, they're inclined to take as much fish as possible, as fast as possible — while it is still there to be taken.

Several attempts have been made at instilling a sense of ownership, and the results have been mixed. In 1982, national economic boundaries were extended by the Law of the Sea Treaty to 200 miles from shore to give countries ownership of the fish within that zone. The idea was that countries would believe they have a stake in preserving their fish. In the case of Canada, it worked. Canada spends about $10 million a year on aerial surveillance and 1,000 armed fisheries inspectors to defend its national catch from fish "pirates."

But most countries cannot afford such extravagant protection. They remain vulnerable to foreign ships that take whatever they can grab, violating quotas and sometimes diminishing future harvests in the process. In fact, even Canada angrily admits that it has not managed to stop all of the overfishing. The Canadian government accuses European Community ships of sitting just beyond the 200-mile Canadian waters and exploiting schools of fish that swim back and forth across the national line, outside of Canadian reach.

So the problem continues, and for the fishermen out of work because of new quotas and shrinking stocks, these are grim times. Canada was able to come up with a massive aid package of about $400 million for the 20,000 fishermen whose livelihoods are threatened by its shutdown of the northern cod fishery. The aid includes retraining, early-retirement incentives, and work-creation projects. But that program, along with the country's ocean-patrol effort, is an exception on the world scene. Most countries find such remedies to be prohibitively expensive. Many of the tens of thousands of fishermen, fish-processors, and merchants who have lost their jobs recently will get only minimal support from governments.

As the need for food continues to increase worldwide, fish are bound to become a more and more coveted resource. A continuing decline in the per-capita fish catch (it dropped nearly 5 percent in 1990) spells trouble for the hundreds of millions of people who depend on fish as a source of protein — and the more than 100 million who make a living either catching, processing, or selling fish. And as the growing demand drives the prices for fish higher, larger shares of the catch are likely to go to countries like Japan and Norway, which can afford to outbid many of the countries from whose waters the fish are taken.

As the world looks to the seas to satisfy its growing appetite, the need for quotas and property rights to protect fish populations becomes increasingly urgent. The United Nations has tentatively scheduled a conference for early 1993 to encourage participants to respect other countries' quotas and to understand the value of setting their own. Past conferences on this subject have drawn little attention, but this one may be different. To those nations that depend on the harvest of the sea, the dangerous divergence of supply and demand may send a signal as bright as any lighthouse warning.

Questions and Concerns

READING 7

1. This author talks about "interconnectedness" – which holds that no part of the natural world is more important than another. He says this challenges two key principles of modern society – "private property and individual freedom." Consider how the rights of private property and individual freedom has led to the collapse of the fishery and how they may have to be curbed if the resource is to survive.

The squid, the cod and who we are

Ray Rogers argues that the decline of Canada's Atlantic fishery is linked to our overall relationship with the natural world.

I WAS at the gas station near my home in Little Harbour, Nova Scotia when Vernon Murphy told me about the time he was fishing out on the edge of Roseway Bank with Hal Harding.

'I was just a kid, maybe 14 or 15,' Vern recalled. 'We had hauled our nets on the way out to the bank, and we got a couple of hundred pounds of herring, enough to bait out five or six tubs of gear.

'By and by, I managed to get a bucket full of squid and I chopped them up and baited nearly a tub of gear on it. We were hauling along on the herring the next morning and we weren't getting a thing. Then we came to the squid I'd baited on and the line was smother white. There was a steak cod as big as a man on every hook as far down as you could see.'

'Hal sunk the gaf into one of those steakers and rolled him up on the rail — its head the size of a wheel rim. He looked back at me with a glint in his eye and said, "It's moments like this, I know who I am."'

❊ ❊ ❊

Canada's Atlantic fishery has changed dramatically since Vern Murphy described that scene from his childhood nearly 40 years ago. Yet the issue of 'who we are' has become even more urgent.

I've lived by 'longline fishing'[1] for the last 12 years. Over that time I've come to see both the fishing industry and the community in which I live threatened. People blame the seals or the foreigners for the decline of fish stocks. But I'd say the cause was much deeper than that: the end of what was one of the world's richest fisheries stems more from how we humans relate to the natural world, to the sea and the creatures that live there.

❊ ❊ ❊

Most people who depend on the sea in Nova Scotia practise small-scale 'inshore' fishing. But the 'dragger fleet' is a different story. These expensive trawlers are packed with sophisticated fish-finding technology. They can travel long distances to reach the fish and they don't need to worry about the weather. They can carry large supplies of food and fuel and tie up in the port nearest the fish. The Canadian 'dragger fleet' now has five times the capacity required to catch its annual quota.

As one of my friends in Little Harbour lamented: 'Ten years ago, I used to swamp load my old boat. Now there's nothing to go for. With these new draggers, all we've done is gear up for disaster. The ocean can't take it. Just look at one of those big draggers come in loaded with 100,000 pounds of twelve-inch long haddock. How long can the stocks take that?'

The noted environmentalist and writer Farley Mowat says this high-tech fishing for profit is 'a financial vortex' which has destroyed the fishery. And I believe he is right. Bigger ships and higher technology take more

Also Worth Reading on ... the Sea

Straddling the line between journalism and literature is Rachel Carson's exuberant classic, **The Sea Around Us**, originally written in 1950. The newest edition *(Oxford University Press, 1989)*, with a solid afterword by marine scientist Jeffrey Levinton, is still a prescient introduction to the history and mysteries of the sea. Another useful journalistic study is **Our Common Shores** *(Earthscan/UNEP, 1990)* by Don Hinrichsen, former editor of the Swedish environmental magazine *Ambio*. Hinrichsen summarizes the major policy and pollution issues in a comprehensive sweep through the world's seas. Also worth a look is **The Living Ocean** by Boyce Thorne-Miller and John Catena *(Friends of the Earth/Island Press, 1991)*, both a survey and a plea for protecting the diversity of marine ecosystems. **World Resources 1992-93** *(Oxford University Press, 1992)* a collaboration between the World Resources Institute, the UN Environment Programme and the UN Development Programme has a fact-filled chapter on oceans and coasts. For an up-to-date activist perspective two journals stand out. **Sea Wind** *(available from Ocean Voice International, 2883 Otterson Drive, Ottawa, Ontario, Canada K1V 7B2)* and **Greenpeace Magazine** *(1436 U Street NW, Washington, DC 20009)*. Write to both for subscription information.

Courtesy of The New Internationalist

investment and that means bigger catches and bigger profits until, as Mowat writes, 'it spirals and you reach the point of no return.'

The earth was not designed to provide stock shareholders with a healthy return on their investment. Trees, fish and whales grow at around five or six per cent a year. In order to compete with the rest of the economy and provide a ten-per-cent return on investment, resource industries like fishing and forestry systematically deplete living natural communities.

Our culture has been taken over by a mindless drive for affluence and growth. In the process we've distorted our relationship with the natural world. The destruction of the fish stocks off Canada's East Coast is just one example of modern society's brinkmanship with environmental limits as we pursue our goal of ever-expanding consumerism.

If the much-ballyhooed concept of 'sustainable development' is to have any real meaning we are going to have to build a new relationship between ourselves and the natural world. And we are going to have to seriously rethink what we mean by 'community': in this sense there is a lot we can learn from the natural world.

Consider the basic ecological principle of 'interconnectedness' which says that no one aspect of a natural community is more important than another. This idea is a direct challenge to two cornerstone beliefs of modern society — private property and individual freedom. And that's what makes many environmental problems so daunting; they go right to the heart of the ways we understand ourselves as a society.

Human beings caught in the web of industrial culture have become a predatory species who have lost a sense of belonging anywhere. The painful process of reacquainting ourselves with the natural world 'in us' and 'out there' must begin soon. Either that or it will be imposed on us. In that sense the plight of villages like my home of Little Harbour is not just a fisheries problem — it's a dramatic sign that our whole culture has gone awry. ∎

Ray Rogers used to fish for a living. He is now a doctoral candidate in environmental studies at York University, Toronto.

---

[1] A fishing method where a length of line from 100 to 200 metres is set with individual baited hooks.

Discussion for The Atlantic Cod Fishery

1. The main protagonists disagree as to why Atlantic Canada's northern cod fishery has collapsed. Don Gillmor summed up their conflicting viewpoints in *Equinox*: "The offshore trawler captains do not think there is a problem, except for seals that eat the inshore cod; the inshore fishermen believe the offshore fishermen are the problem; the plant workers think it's foreign overfishing; the foreign experts say its Canadian mismanagement; the provincial government blames the federal government; the federal government takes the visionary position that too many fishermen are chasing too few fish; an independent study suggests a serious resource problem exists, one that stems partly from inaccurate Department of Fisheries and Oceans (DFO) estimates of the cod stocks for the past 13 years; the scientists who made the estimates deny there is a resource problem; and a woman at the Petro-Canada station at Goobies, Newfoundland, believes it's God's will." Do you think there is truth in all these arguments, or are some more likely than others?

2. Fishing technology has improved dramatically in recent years. Larger ships are operated by smaller crews; the remaining fish populations are tracked with radar and captured with huge bottom-dragging nets. This kind of super-efficiency is in keeping with trends in forestry (clearcutting and giant pulping mills) and agriculture (mechanized farming). But is it a good thing? What happens to the environment? "Like the tree harvester and the dynamite stick," writes Silver Donald Cameron, "the dragger applies raw force to a complex and sensitive ecological system." Furthermore, what happens to the people put out of work by mechanization? What happens to communities that depend on fishing? Is it realistic to call for a ban on huge factory trawlers so that fishing can once again become a more labour-intensive, smaller-scale enterprise?

3. Canadian federal fishery scientists made predictions about cod populations based on inaccurate data. The Total Allowable Catch set annually by the Department of Fisheries and Oceans was based on these predictions. How could the scientists be so wrong? Do you think there could have been political motives for overinflating the figures? What would these be? Can you think of other environmental issues where science and politics have collided and the result has been continued destruction of a resource?

4. The Canadian government has tried to curtail foreign fishing through diplomatic channels, but ships from the European Community (EC) continue to take far more fish than stipulated by the Northwest Atlantic Fishery Organization (NAFO). In 1986, for example, NAFO set the EC quota at 25,665 tonnes; the fishermen took 172,163 tonnes. Canada is now spending millions each year to patrol its fishing grounds and apprehend European offenders fishing illegally within the 200-mile limit. The Minister of Fisheries and Oceans is talking ominously about taking "unilateral measures" against foreigners ignoring NAFO quotas. Could the desecration of an ecological resource such as Canada's East Coast cod fishery conceivably lead to international economic sanctions or even war? Can you think of other environmental problems in the world where one nation is so victimized by others that hostilities could result? Could environmental protection be considered a national security issue?

5. After haddock fishing was regulated within Canadian fishing boundaries in 1985, the haddock stocks made a comeback. Today, there are more haddock and they are larger in size. It appears as if Canada's Atlantic haddock fishery is headed toward achieving sustainable yield. How could this be achieved with the northern cod stocks? What strategies would be necessary?

Introduction

Arctic Pollution

This issue examines the impact of environmental contamination on Arctic ecosystems and native peoples. The Far North has always been considered unspoiled and pristine. Recent research shows it is not only becoming polluted, but that the Arctic is especially vulnerable to pollution because of its climate and extreme latitude.

"The North is the only place where Nature can still claim to rule, the only place as yet but little vexed by man. All over the globe there spread his noisy failures; the North alone is silent and at peace. Give man time and he will spoil that too."

Stephen Leacock, 1937.

When Stephen Leacock, the famous Canadian humourist, wrote these words in 1937, the Canadian Arctic was indeed still unspoiled. Mining and oil extraction were insignificant; bulldozers and snowmobiles were nonexistent; native people were largely unaffected by industrialized western culture; and the first signs of widespread global pollution from the petrochemical industry were still 20 years in the future. Today, the Arctic is burdened by all these problems, and as Leacock predicted we have begun to spoil its unique environment.

These are the key issues affecting the environment of the Canadian Arctic:

1) **Vulnerability** – The Arctic is more vulnerable to pollution because of its extremely cold climate, its lack of sunlight, and its position at the "top" of the world where pollution is deposited by several large rivers and prevailing ocean and atmospheric currents. Ultraviolet solar radiation and heat help degrade and evaporate pollutants in more southern regions. In the Far North, where winter sunlight is limited and weak and the air temperature is seldom warm, contaminants tend to resist degradation and evaporation. This leads to "biomagnification" or "bioaccumulation" – an environmental buildup of chemically-stable compounds such as polychlorinated biphenyls (PCBs), organochlorines (like dioxin), and certain pesticides (like DDT and chlordane).

Once in the environment, these chemicals are consumed by animals, birds, and fish, concentrating further each time they move up the food chain. For example, PCBs in sea water accumulate in the fatty tissues of zooplankton, where they concentrate by a factor of two million. The zooplankton are consumed by fish like the Arctic cod, further concentrating the PCBs. The cod are, in turn, eaten by seals and beluga whales. At the end of the chain are polar bears, who eat seals and have levels of PCBs in their livers three billion times as concentrated as in sea water. Inuit people who consume wild game also have unacceptably high concentrations of PCBs and radioactive particles called "radionuclides" in their blood and breast milk.

2) **Local Waste** – Northern communities are faced with the basic dilemma of having to dispose of trash and sewage in an environment that doesn't readily break down these wastes. Consequently, all waste tends to accumulate in and on top of the frozen ground (known as "permafrost"). There are also many abandoned industrial sites throughout the Arctic, including mining sites, tailings ponds, and old U.S. military radar installations. Chemical wastes left at these sites often contaminated soil and groundwater; piles of rusting barrels and discarded machinery are an aesthetic blight.

3) **Global Pollution** – The Arctic is a bellwether for global pollution. Since the early 1960s, scientists have been detecting PCBs and toxic pesticides such as DDT and hexachlorocyclohexane carried to the Arctic from the south on prevailing air currents. While controls on DDT and PCBs in Canada, the United States, and other western countries have resulted in declining levels of those particular chemicals, the presence of other powerful pesticides and organochlorines from industrial incinerators is growing. (although DDT is still being carried to the Arctic from as far away as Southeast Asia, where it is still used to control insects.)

"Arctic haze" is another problem. Each spring since the 1960s, industrial soot, hydrocarbons, and sulphates gather in clouds of this smoggy haze that darken snow and taint the air.

Russia and other former Soviet countries contribute pollution from inefficient factories, coal-fired power plants, and through three rivers (the Yenisey, Lena, and Ob) that account for 70% of all river drainage into the Arctic Ocean. These rivers contain high levels of organochlorines and other toxic industrial wastes.

One particularly serious form of global pollution gathering in the Arctic – the accumulation of chlorine- and bromine-based chemicals in the stratosphere – has eroded the Earth's protective ozone layer above the northern hemisphere. In the winter of 1992, scientists at the National Aeronautics and Space Administration (NASA) in Washington, DC found record levels of ozone layer depletion. The problem, they say, has become even more pronounced than anticipated and is now cause for alarm. It is expected that a hole similar to the one over Antarctica will open above the Canadian Arctic in the next few years. This will result in increased human skin cancers and eye cataracts and will harm animals and crops exposed to the higher levels of ultraviolet radiation.

4) Radioactive Contamination – Radioactivity – in the air, on land, and in water – was highest in the Arctic during the 1960s when the United States and Soviet Union conducted atmospheric nuclear weapons testing. It has declined steadily since then, with the exception of 1986 when the Chernobyl reactor meltdown in the former Soviet Union caused a worldwide elevation of ambient radioactivity. A grave new threat is posed by nuclear waste and used nuclear reactors that were dumped for years by the Soviet military off Siberia's Arctic and Pacific coastlines.

5) International Response – There is a general consensus among Arctic nations that the only way to prevent the continued build up of pollutants in the Arctic is through international controls. Canada is a signatory on more than a dozen bilateral and multilateral agreements aimed at protecting the Arctic environment. Recently, eight countries within Arctic regions – Canada, the United States, Norway, Finland, Sweden, Russia, Iceland, and Greenland – have been conducting ongoing multilateral consultations about Arctic pollution through the Finnish Circumpolar Initiative on Protection of the Arctic Environment. But because the problem is global in origin, these eight countries have little power to prevent long-distance pollution.

6) Canadian Response – Through its Green Plan, the Government of Canada has committed $100 million over six years (1992 to 1998) to The Arctic Environmental Strategy. The money will be used to set up a network of stations to monitor water pollution, to cleanup hazardous and non-hazardous waste dumps, and to conduct research and monitoring programs related to the long-distance transport of contaminants, and to determine the impact of pollution on Arctic ecosystems and human health. The program is also intended to assist northern communities in reconciling their economic and environmental priorities.

The government of the Northwest Territories is the first in Canada to allow individuals to take legal action on behalf of the environment. Under the Northwest Territories Environmental Rights Act, any two people can request government action against polluters.

7) Sustainable Development – Resource companies in the oil and mining industries have for many years polluted the Arctic and contaminated the land on which they operate. Since the increase of environmental awareness throughout North America in the 1980s, many of the larger resource companies have begun to make "sustainable development" part of their operating codes. The concept calls for development of land and resources that does not harm the environment, and for the fostering of consensus among "stakeholders" (including native groups) about the benefits and sacrifices involved in a project. Companies must also operate under environmental protection laws which have tightened considerably over the years.

8) Climate Change – Scientists can't be sure about the way global warming will affect the Arctic. But there are two obvious concerns. One is ice. Increased temperatures could melt the fantastic volumes of Arctic ice, causing increases in world sea levels. A greater concern in the short term is the well-being of Arctic plant and animal species. Biological productivity is slower in the Far North, and many species rely on short "windows of climatic opportunity" during the spring and summer for breeding and feeding. Climatic disruptions could close these windows and conceivably wipe out entire populations.

The probability of increased ultraviolet penetration due to the growing hole in the polar ozone layer could affect the reproductive abilities of zooplankton, an extremely important nutrient resource at the base of the Arctic food chain.

Reading 1

Questions and Concerns

READINGS 1 THROUGH 4

1. Give three reasons that pollution tends to have a greater impact on the Arctic than on more southern regions of the world.

2. The pesticide DDT was banned in North America in the 1970s. Why is it still showing up in the flesh of Arctic animals and the milk of Inuit mothers?

3. Why are large resource corporations improving their approach to the Arctic environment?

The not-so-pristine Arctic

From plankton to polar bears, the food chain is contaminated by global pollution

By Karen Twitchell

THE HAMLET OF Broughton Island, off the east coast of Baffin Island, is about as far removed from the smokestacks and sewers of industrial society as you can get. Situated some 2,700 kilometres from the urban sprawl of southern Ontario and 4,000 kilometres from the industrial heartland of Europe, it seems unspoiled. The Inuit there live largely off the sea, hunting whales, seals and other wildlife.

But the remoteness of Broughton Island has not granted it, or the rest of the Canadian Arctic, immunity from pollution. In the last 20 years, scientists have detected an increasing variety of toxic contaminants in the North — pesticides from agriculture, chemicals and heavy metals from industry, and even radioactive fall-out from Chernobyl. They have been found in snow, ice, water and air, and have contaminated every level of the arctic food chain from plankton to polar bears.

These same substances have pervaded ecosystems virtually everywhere, but they are especially worrisome in the Arctic, where reduced sunlight, extensive ice cover and cold slow the breakdown of chemicals. Moreover, Inuit and other northerners eat large quantities of foods from wildlife near the top of the food chain, a diet that likely exposes them to higher levels of contaminants than other Canadians.

Some heavy metals and organochlorine compounds, such as polychlorinated biphenyls (PCBs) and pesticides, increase in concentration and toxicity as they move up the food chain. "The concentration of PCBs in polar bears is up to three billion times higher than in sea water," says Derek Muir, a research scientist with the Department of Fisheries and Oceans. Organochlorines pose an additional concern since they accumulate in whale and seal blubber and caribou fat, considered delicacies by the Inuit.

Scientists first discovered PCBs and the insecticide DDT, both linked to

Courtesy of Karen Twitchell

reproductive failure and birth defects, in arctic wildlife in the late 1960s. Since then, a plethora of man-made chemicals and industrial pollutants has been found throughout the food chain, partly due to improvements in measuring devices. "The analysis we can do now is a thousand times more sophisticated than what was done in the early 1970s," says Ross Norstrom, head of chemistry research for the Canadian Wildlife Service. Today scientists can measure substances in parts per quadrillion — equivalent to detecting a two dollar bill in an area the size of Canada.

At first, arctic contamination was largely blamed on chemical leaks at abandoned DEW Line stations. But in 1985, a survey concluded that the impact of these leaks was likely "small and localized." The concensus now is that pollutants from around the world are being carried north by rivers, ocean currents and atmospheric circulation.

The international scope of this problem has heightened concern among many northerners, says Ron Mongeau, a member of the Baffin Regional Health Board in Iqaluit. "They view it as a situation that's out of their control. They aren't causing the pollution, they have no way to stop it, and yet it directly affects them."

Broughton Islanders became acutely aware of the problem in the mid-1980s when scientists carried out extensive studies to determine the health risks of PCBs in local food. The hamlet was chosen because the per capita consumption of "country food" there is thought to be the highest on Baffin Island. The surveys indicated that 13 percent of those who eat a traditional diet ingest PCBs above the tolerable daily intake level set by Health and Welfare Canada — one microgram per kilogram of body weight. The survey also revealed that 90 percent of the community's PCB intake came from four foods: caribou, narwhal, seal and walrus.

Despite these findings, the study revealed no evidence of health problems that could be directly related to toxins in country food. Broughton Islanders were told they should continue to eat wildlife because the benefits outweigh the risks. Harriet Kuhnlein, a nutritionist at McGill University involved in the study, says traditional food is richer in nutrients and far cheaper than store-bought food. It is also an integral part of the Inuit way of life. Still Kuhnlein admits, the risk of exposure to PCBs and other contaminants is "largely unknown."

So far Broughton Island is the only northern community that has been placed under the microscope, and the research there focused on PCBs. Little is known about the levels of contaminants that Inuit in other communities are exposed to.

How specific contaminants reach the Arctic varies from substance to substance. PCBs and pesticides, such as DDT, toxaphene and chlordane, are chemically stable and resistant to biological degradation. After PCBs are spilled, or pesticides sprayed on crops, they evaporate and are carried long distances by prevailing winds. Pesticide-laden soils and organochlorines from industrial incinerators are also transported this way. Once deposited on land or water, they may re-evaporate and be on the move again.

The veil of smog called arctic haze that occurs in the North from December to April is another carrier of toxic substances. In winter, the prevalent air currents are from the east, causing an influx of pollutants, including dioxins, from incinerators, pulp and paper mills, metal recycling plants and other industries in Europe and the Soviet Union. Muir and Norstrom have found that levels of dioxins in polar bears and ringed seals in the central Arctic exceed amounts in wildlife farther south — strong evidence that the pollutant originated somewhere other than North America. The most toxic dioxin, 2,3,7,8 TCDD, was among the substances they detected.

The Yenisey, Lena and Ob rivers in the Soviet Union, which account for almost 70 percent of river drainage into the Arctic Ocean, may carry significant quantities of organochlorines, according to a recent federal report on the arctic environment. It also suggests that the impact of spring run-off could be critical in coastal waters because it coincides with the growth period of marine life.

Little research has been done on the amount of organochlorines in Canada's northern watershed. But in a 1987 study of rivers draining into Hudson and Ungava bays, the insecticide hexachlorocyclohexane (HCH) was detected at levels just below the maximum recommended by Canadian guidelines. The same year, Barry Hargrave of the Bedford Institute of Oceanography in Dartmouth, N.S., found that hexachlorobenzene (HCB), a by-product of pesticide manufacturing, and HCH were the major contaminants in surface sea water near Axel Heiberg Island. Both substances are known to cause cancer in lab animals.

Scientists have also been trying to determine whether specific contaminants are increasing in the Arctic or not. Early evidence suggests that international regulations on some substances have helped stem the flow northward.

Richard Addison, a research scientist with the Bedford Institute, found that PCB levels in arctic ringed seals declined by 50 percent between 1972 and 1981. He attributes this to controls placed on PCBs in the 1970s by Canada, the United States and other western countries.

Addison is now analysing samples of ringed seal tissue collected in the western Arctic in 1989. "I expect the levels in ringed seals will be lower again, but I don't think the decline will be as steep as during the 1970s," he says.

PCBs consist of up to 209 different chemical components (or isomers), some of which degrade more readily than others. How quickly a particular PCB component breaks down depends on its molecular structure. Less persistent PCBs may account for the big decline in levels already seen in the Arctic. Addison predicts the more highly chlorinated PCBs may take years, if not decades, to disappear. Much of the estimated 370,000 tonnes of PCBs released into the environment since 1930 continue to circulate around the world.

DDT, also banned in North America during the 1970s, declined three to four times faster than PCBs in East Coast seals, yet its levels remained fairly constant in arctic ringed seals. One reason for its persistence in the North may be that it is still widely used in Southeast Asia and the Middle East, says Addison. Air currents may be carrying DDT across the northwest Pacific, depositing it in the Arctic.

It is too soon to tell whether substances discovered more recently in the North are on the increase or not. "We need more information on trends, especially concerning toxaphene," says Muir.

In 1987, Muir and his colleagues found higher levels of toxaphene than PCBs in arctic whales and fish. In fact, its concentration in fish was higher than any other organochlorine.

A complex mixture of up to 200 substances, toxaphene can be lethal to fish and has been used to rid lakes of undesirable species. Its main use, however, was as an insecticide on cotton in the south-central United States, until it was banned in the 1980s.

"Twenty to 30 percent of the amount sprayed would end up in the atmosphere by evaporating off the ground," says Muir. Prevailing winds then carried it north.

Toxaphene is still sprayed on crops in India and China. "The fact that there isn't a complete ban everywhere means it will continue to be put into the environment and circulate globally," says Muir.

In the meantime, scientists are trying to determine if contaminants already in the Arctic pose an immediate danger to wildlife. So far there is no evidence of harmful effects, likely because the pollution levels are generally lower than farther south, where organochlorines have been linked to reproductive failure in marine mammals. Belugas in the St. Lawrence River, for example, have 25 times more PCBs in their tissues than arctic belugas and half as many young. Research during the 1970s and early 1980s also linked PCBs to low birth rates among ringed seals in the Baltic Sea, where concentrations were 100 times greater than in seals in the Canadian Arctic.

Yet despite lower levels in Canada's North, there is concern over the amount of organochlorines the young of some species are exposed to. Polar bear cubs, which nurse for up to two years, may be getting a substantial dose in their mother's milk. During that time, the female may not eat anything for up to four months, causing her fat reserves to decrease and the concentration of organochlorines to rise, says Norstrom. These contaminants eventually end up in her milk. "We only have one analysis of polar bear milk so far, but indeed the concentration of PCBs and DDT were very high."

A 1988 study also found that the level of PCBs in human milk fat was five times greater in samples from Inuit women in northern Quebec than from Caucasian women farther south. The higher concentrations "are clearly because of dietary habits, specifically consumption of marine mammals," says Eric Dewailly, head of the study and of Environmental Health Services at the Laval University Hospital in Ste-Foy, Que.

Dewailly and others suspect that exposure to chemicals such as PCBs may lower immunity. Last year, he began research to determine whether higher rates of infectious disease among Inuit children in Quebec are due to toxic contaminants in breast milk.

Other scientists have linked high PCB exposures to a severe form of acne, numbness of limbs, decreased birth weights and smaller heads in newborns. Recent studies in Michigan have also found a link between prenatal exposure to PCBs and slight reductions in mental development.

It will take considerably more time and research to determine whether long-term exposure to lower concentrations in country food is detrimental to human health.

"We only know a little bit about levels (of contaminants) in some arctic species," says Andy Gilman, a toxicologist at Health and Welfare Canada. "And in terms of suspecting impacts, we can only extrapolate to some animal studies and a few human poisonings and say that at these levels we think these type of things might occur. But the data just doesn't exist now."

Scientists do not know at which concentration PCBs or other substances begin to cause reproductive failure. Moreover, most contaminants do not occur in isolation. The combined effect of chemicals reacting with each other in the environment is unknown.

Another problem is that not all PCBs are created equal — some are more toxic than others. "We're now realizing, as science becomes more sophisticated, that it does make a difference which of the hundreds of PCBs are present in a mixture," says Gilman. "That creates an entirely new level of complexity. Suddenly, regulations based on total PCBs may in some cases be too protective and in others inadequately protective."

The federal government intends to spend another $30 million over the next five years on more research. In the long run, though, everyone agrees the only way to prevent industrial contaminants from polluting the Arctic is international controls.

One of the research goals is to pinpoint the sources of pollutants and whether their atmospheric concentrations are increasing or not, says Leonard Barrie, senior scientist with the Atmospheric Environment Service. "We are now proposing ongoing measurements of organochlorines in the

High Arctic to find out how much is in the air and how it varies seasonally."

Scientists also hope to learn more about the pathways of contaminants entering the Arctic. Since the obvious solution is to control pollutants at the source, researchers must first understand the complexities of atmospheric circulation to convince polluters their actions have global implications, says Barrie. "That's the reason for doing research, but research in itself is not the entire answer."

It will take considerable political pressure to stem the flow of organochlorine pesticides and other toxic substances that continue to pour into the global environment. In January, representatives from Canada, the United States, Sweden, Norway, Finland, Iceland, Greenland and the Soviet Union met for the third time to develop an arctic environmental protection strategy. The goal is to share research and eventually push for widespread controls on the use of pesticides worldwide.

The United Nations Environment Program in Geneva has also established the International Register of Potentially Toxic Chemicals to gather information on the international use of pesticides. But it is a difficult task confounded by governments that do not collect statistics or register substances, says Garth Bangay, chairman of the federal government's committee on toxic contaminants in the Arctic.

In the meantime, scientists and northerners express frustration at the questions still unanswered and the uncertainty of long-term effects.

"We definitely need more information," says Tagak Curley of Rankin Inlet, a former minister in the Government of the Northwest Territories. "They only studied one community (Broughton Island). What about the rest of the people in the central Hudson Bay region, Quebec and other areas? To what extent is the contamination in wildlife there?"

For now the Inuit of Rankin Inlet and other communities across the North continue to harvest their food from a land polluted by the industrial world — a reminder that no place on earth can escape the impact of our collective assault on the environment.

Karen Twitchell is assistant editor of Canadian Geographic.

Northern Perspectives

Volume 18, Number 3, September-October 1990
Published by the Canadian Arctic Resources Committee

Second Class Mail
Registration No. 5366

ARCTIC POLLUTION:

HOW MUCH IS TOO MUCH?

*In recent years, as environmental concerns
have percolated toward the top of the policy agenda,
Canadians have discovered that all is not well
in the Great White North. Our frozen slice
of Eden has been thrust into the 20th century,
and with that displacement has come
a disquieting array of very "un-arctic" problems.*

(cont'd on pg 2)

Originally appeared in Northern Perspectives, published by the Canadian Arctic Resources Committee.

To portray the Arctic as little more than a chemical dumpsite would be misleading. Compared with other regions on the face of the globe, the Arctic remains a pristine wilderness. And yet the lexicon of "arctic" pollution is no less fearful than that of industrial regions in North America or Europe. Furans, cadmium, dioxins, chlordane, selenium, PCBs, mercury, radioactive fallout—all are now a part of what has long been regarded as the planet's most vulnerable and fragile ecosystem.

"Arctic contaminants" is a deceptive phrase. The Arctic is polluted because the world in which we live is polluted. What is disturbing is that the "arctic wastes" of legend are now taking on a new meaning; remoteness and the absence of indigenous pollution sources no longer guarantee the well-being of northern communities and the viability of wildlife populations.

It is indeed a sad irony that the traditional, land-based economy of northern aboriginal populations, based on an abiding respect for natural ecosystems, is threatened most by the effects of non-indigenous pollution.

But while pollution is pollution no matter where it exists, the Arctic does present a potentially more serious situation. Detection, monitoring, and clean-up are difficult due to climatic conditions, remoteness, and the shifting interplay between land and sea-ice. Whereas solar radiation generally speeds the break-down of contaminants, the reduced level of sunlight in the Arctic lengthens the degradation process and increases the likelihood that toxic substances will find their way into the food chain.

A recent report prepared by the Department of Indian Affairs and Northern Development notes:

"In making comparisons with other ecosystems, it is, however, important to stress that the lower concentrations detected in the arctic do not diminish the importance of their effects on ecosystem health."

In addition, the report found that "there are currently insufficient data on sources, sinks, and path-

> ...the federal government's "Green Plan" strategy is a step in the right direction, but turning tough talk and noble intentions into implementation and enforcement may prove a difficult task...

ways and spatial and temporal trends of contaminants in the arctic ecosystem."

In many respects, the Arctic serves as a benchmark for global pollution. The spillover of industrial contaminants from other regions via air, ocean, and river currents tells us a great deal about the overall health of the planet. And although new, often challenging, questions have been posed, the mere fact of pollution in the Arctic has sounded the alarm and helped rouse policy makers to the need for concerted action.

There are no easy answers to arctic pollution. The bioaccumulative effect associated with repeated exposure to toxic substances counters the natural defences of most organisms and means potentially dangerous levels of toxins can span generations. And while exposure to toxic pollutants is rarely fatal in humans, the effects—both direct and indirect—can be debilitating.

On the domestic front, the federal government's "Green Plan" strategy is a step in the right direction, but turning tough talk and noble intentions into implementation and enforcement may prove a difficult task as the spectre of economic recession looms on the horizon. The Green Plan's northern component, the Arctic Environmental Strategy, is as yet little more than a collection of finely tuned phrases designed for public consumption.

In the international North, several recent initiatives display much promise—the ongoing multilateral consultations known as the "Finnish Initiative," the establishment of the "Northern Forum" as a permanent decision-making organization of regional leaders, creation of the International Arctic Science Committee—but each must recognize the difficulty inherent in addressing a global problem on a regional basis.

Efforts to deal with the problems of industrial pollution at the multi-lateral level involve a range of geopolitical considerations and must go well beyond the purview of the circumpolar "eight"—Canada, the United States, the U.S.S.R., Iceland, Denmark, Sweden, Norway, and Finland. Studies of arctic pollution point to heavy industry in eastern Europe and neighbouring regions as primary sources, but traces of pesticides and related chemicals from as far away as southeast Asia have been detected in the tissues of arctic wildlife.

What is required, then, is agreement among the arctic states on the scope of pollution problems in northern regions as well as a political mechanism by which the eight can deal effectively with non-arctic polluters.

The concept of an arctic basin council, first proposed in the 1988 CARC report, *The North and Canada's International Relations,* is now being actively explored. CARC, with the support of the Walter and Duncan Gordon Charitable Foundation, has convened a panel of northern native leaders and researchers to prepare a framework document for release early in 1991.

Later this year, CARC, in cooperation with the National Capital Branch of the Canadian Institute of International Affairs, will publish *The Arctic Environment and Canada's International Relations,* a follow-up volume to the 1988 report which will review the state of Canada's arctic environment and offer a number of recommendations. The section which follows has been adapted from material prepared for that volume.

Pollution in the Arctic

Due to its relatively small human population and the general absence of large-scale industrialization, the arctic region does not represent a major source of pollution. Yet, in nearly all parts of the circumpolar Arctic, increasing pollution of the natural environment is a serious problem. In most places, current levels of pollution are low—considerably lower than in most urban and industrialized areas in the mid-latitudes—but still cause for concern because:

- contamination of snow, waters, and organisms with "imported" pollutants is a phenomenon of the past few decades and appears to be increasing;

- arctic ecosystems show indications of being much more susceptible to biological damage at low levels of pollutants than higher-energy ecosystems in temperate latitudes; and

- many arctic organisms, adapted to storing biological energy, become accumulators and concentrators of organic pollutants and toxic metals, so that animals at the top of the food chain, including humans who eat local foods, may carry pollutant concentrations much higher than levels in the ambient environment.

In addition, because the effects of pollution from distant sources are in some cases clearly discernible in the Arctic, arctic regions serve as important indicators of environmental changes in the planet as a whole—and of the need for international action to control further deterioration.

Pollution of international significance is generated within the arctic regions in only a few areas. The most important of these at present are the highly industrialized Kola Peninsula and White Sea regions of northwestern Russia, and the large metallurgical and wood processing complexes of north-central Siberia. Each of these areas contributes to circumpolar pollution; the prevailing winds carry airborne pollutants over the central Arctic Basin, and the rivers deliver their contaminants to the Arctic Ocean.

Other potential arctic sources of pollution are the areas of present or potential hydrocarbon production and transport in the arctic Soviet Union, arctic Canada and Alaska, and the adjacent seas. The possibility of damage to sensitive arctic marine ecosystems by routine or accidental oil spills or disposal of radioactive waste is a serious concern for indigenous people and others who make use of, or value, the Arctic's marine and aquatic resources.

Transport of Pollutants

Pollution from lower latitudes is carried into the Arctic by atmospheric circulation and ocean currents. Global atmospheric circulation patterns are such that eastward-moving air masses in northern mid-latitudes become polluted near the surface and then may get carried at moderate or higher elevations to the arctic regions, where they descend and may deposit their impurities. The main pathways by which airborne pollutants reach the Arctic are over northern Europe and Asia and then across the Arctic Ocean to northern Canada and Alaska, although excursions of polluted air from the industrialized midwest

Canada and the Arctic Ocean are not uncommon.

The rapid transport of radioactive contaminants from Chernobyl in the southwestern U.S.S.R. to northern Scandinavia, and its incorporation into vegetation and the flesh of reindeer, is a recent unfortunate but convincing demonstration of the effectiveness of transport of pollutants from southern latitudes into the Arctic. Careful analysis of impurities in snow from various parts of arctic Canada, Alaska, and Spitzbergen has shown that some of the chemicals whose origin can be identified with reasonable confidence have come from industrial sources in western Europe and the western Soviet Union, or from agricultural chemicals typically used in India and Southeast Asia.

By the marine route, pollutants reach the Arctic through one major point of entry, the northeast Atlantic. Chemically stable or slow-reacting pollutants from industrialized eastern North America are carried by winds or rivers into the Atlantic Ocean and then northward by the Gulf Stream and North Atlantic Drift into the Arctic Ocean. Augmented by drainage and winds from Europe and by north-flowing Siberian rivers, they are carried under the arctic ice where they remain protected from sunlight and vigorous oxygenation which otherwise would hasten their chemical break-down. By these means, much of the far-travelled and persistent waste products of the industrialized world appear to be ultimately deposited in arctic regions. A portion of these, to which may be added materials deposited directly from the atmosphere, becomes incorporated in the upper layers of Arctic Ocean waters and returned to the Northwest Atlantic, where they sink to lower ocean depths and are carried slowly southward as the Atlantic Deep Water Current, eventually spreading at depth throughout the World Ocean.

Arctic Haze

About 1950, a thin but persistent brown haze became apparent in the previously very clear arctic skies of northern Alaska and the western Canadian Arctic in late winter and early spring. The phenomenon increased and became more widespread, and was dubbed "arctic haze." The haze is now a frequent but varying springtime feature from north Greenland across the arctic coasts of Canada and Alaska, and is occasionally seen in eastern Siberia.

Although the absolute concentration of pollutants in arctic haze today is less than that of city smog, levels of suspended particles in the atmosphere in northernmost Canada have been found to be 20 to 40 times as high in winter as in summer. Chemical analysis has shown that the haze consists of particles of largely industrial origin, mostly soot, hydrocarbons, and sulphates. Studies suggest that it commonly originates in heavily polluted air that contains large amounts of soot or carbon whose energy-absorbing properties allow discrete air masses to remain aloft in the very cold arctic atmosphere and travel tens of thousands of kilometres across the Arctic Ocean at the edge of the zone of advancing spring sunlight. Upon reaching the North American Arctic, the haze particles appear to be deposited, the carbon darkening the snow slightly and hastening melting, the sulphates adding a little acidity to the already naturally acid waters, and the small but possibly significant amount of toxic metals adding to the pollutant load in these areas.

From present evidence, it does not appear that arctic haze has a significant deleterious effect on the environment, except that the warming of the air through sunlight absorption by the haze may directly affect weather patterns in the northern hemisphere. But the phenomenon bears careful watching, and is a conspicuous example of how a variety of activities in different countries can affect environmental quality in distant parts of the globe.

Stratospheric Ozone

The layer in the upper atmosphere that is slightly enriched in ozone (O_3) and which acts as a filter to prevent damaging ultra-violet and other solar radiation from reaching the earth's surface in concentrations greater than that to which living organisms have been accustomed has been thinning over most parts of the planet over the last two decades—or as long as accurate measurements have been available.

Over southern Canada, the decrease has been three or four per cent. During the past 10 years, however, dramatic decreases—as much as 50 per cent—have been observed over Antarctica for a

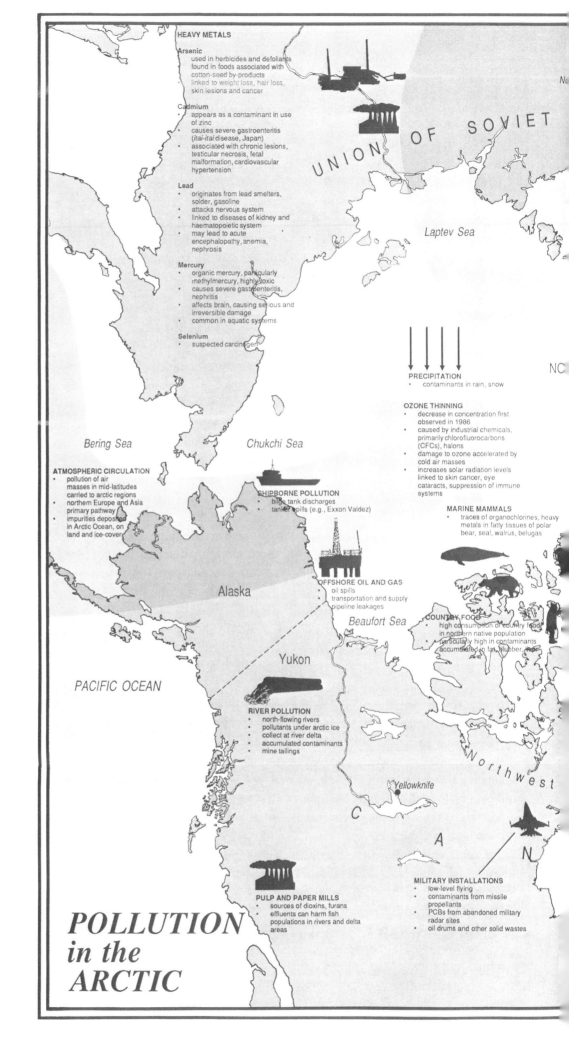

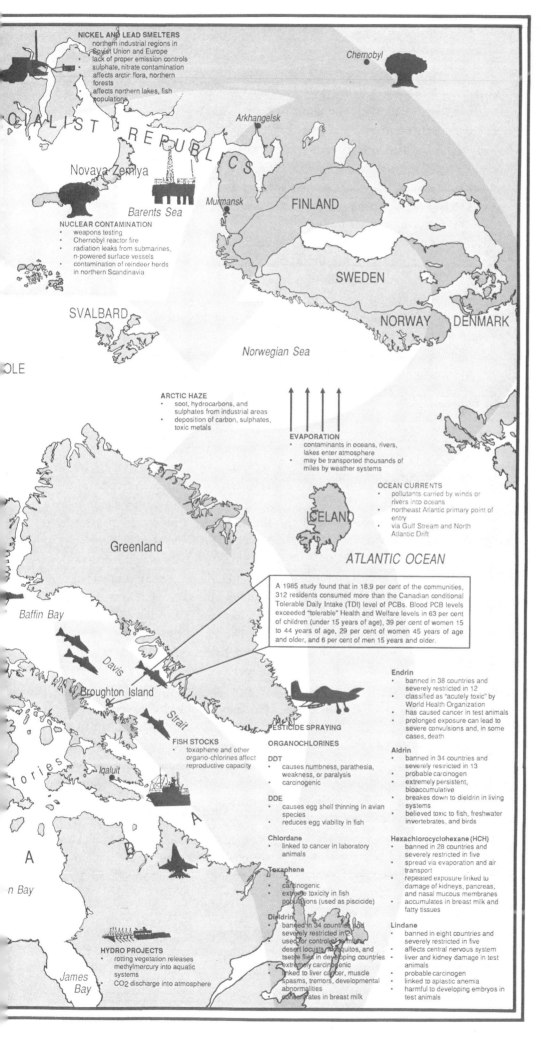

period of about two months in late winter— the so-called Antarctic ozone hole. In 1986, a similar but smaller area of decreased ozone concentration was discovered in the Arctic. No depletion was observed in 1987 or 1988, but the "hole" reappeared in 1989. Some of these variations may be due to natural causes, but careful research in both polar regions shows that much of the change must be due to industrial chemicals, mainly chlorofluorocarbons (CFCs), together with related compounds known as halons, and other "ozone-depleting" chemicals widely used all over the globe. The ozone decrease is greatest in the polar regions because in intensely cold air masses (below -85°C), ice clouds form in the ozone layer, providing nuclei that facilitate the chemical reactions that destroy ozone. The increased solar radiation that can reach the surface through a depleted ozone layer can cause skin cancer, eye cataracts in humans and other animals; suppress immune system response; damage shallow-dwelling marine organisms; and inhibit germination of seeds. The potential seriousness of these effects is such that an international agreement to control substances that may damage the ozone layer was signed in Vienna in 1985. This was followed by a protocol drawn up in Montreal in 1988 to reduce use of CFCs by 50 per cent by 1999, and further strengthened by an agreement reached in Helsinki in 1989. Most industrialized countries have now ratified the Montreal protocol, but it is already apparent that the provisions are not stringent or comprehensive enough, and several countries, including Canada, have made commitments to cease manufacture and phase out use of CFCs and other ozone-destroying chemicals by the end of the present century.

The depletion of ozone over the Arctic in late winter itself probably has minor environmental or biological effects, but it is a dramatic symbol and important monitor of a global environmental trend and a potentially serious biological, economic and health problem. The arctic ozone depletion therefore has great significance for the world environment, and for human health and the economy.

Toxic Substances in the Arctic

The accumulation of toxic substances in the arctic terrestrial, aquatic, and marine ecosystems represents a potentially serious threat to the regional environment. Toxic compounds, particularly organochlorines and some heavy metals, have been found in potentially worrying amounts in snow, waters, and organisms in arctic North America, Greenland, and Svalbard. The organochlorines (e.g., dioxins, furans, PCBs) accumulate in fatty tissue or bone marrow; and because arctic animals consume, and develop, considerable fatty tissue in order to conserve heat, and depend on reserves of marrow during periods of inactivity or hibernation, some remarkably high concentrations of toxins can result through short and simple food chains.

Toxic metals are also a worry. Seals in some parts of the Arctic carry large body burdens of mercury and cadmium, apparently, but not certainly, from natural sources, and this contamination is passed to bears and humans. The liver, fatty tissue, and milk of polar bears and whales from many parts of the Arctic have shown an increased level of chemicals from modern pesticides and herbicides over the past three decades, although comparisons with the past are not very reliable because earlier sampling and analytical techniques were not as accurate as those of today. There is evidence that since 1986 in some arctic mammals the levels of certain polychlorinated hydrocarbons whose use in North America is now regulated have decreased, suggesting that the arctic environment responds quite rapidly to environmental control practices in lower latitudes. Such changes further confirm the sensitivity of arctic ecosystems to pollution, even from distant sources—a change in either direction has an effect on the Arctic.

The human physiological response to the accumulation of toxic materials in the arctic environment is a cause for concern. Northern residents eat a higher proportion of "country food," particularly fatty meats, than most of their counterparts in lower latitudes. Because the fatty meats and organs of both marine and terrestrial animals in the Arctic are major concentrators of organochlorines and heavy metals, serious amounts of toxic contaminants may be accumulated by human populations.

Careful study of the diets and chemical physiology of residents in arctic communities has shown that persons whose diet includes a high percentage of local meat have, in fact, a higher level of identifiable "chemicals used only in the South" (i.e., from agricultural pesticides) in their body tissues and mothers' milk than those who import most of their food from the South. In Canada, this situation is not yet serious; but it needs to be watched carefully.

Radioactivity in the Arctic

The problem of radioactive contamination of the arctic environment raises atmospheric, biological, and human health issues. The arctic regions, especially Alaska, northern Canada, and Greenland, received measurable amounts of radioactive fallout from the atmospheric testing of nuclear weapons in the 1960s and early 1970s. This radioactivity is still present in the slow-growing lichens and mosses upon which caribou and reindeer feed, and adult Inuit from the affected regions carry body burdens of radioactivity higher than the North American average. The comparatively small amount of radioactivity that was released to the environment in the 1986 nuclear accident at Chernobyl was deposited over a wide area; but the effect of the fallout was relatively greatest in northern Scandinavia, where the meat of animals who grazed on the affected vegetation was declared unfit for human consumption and the effect on the local economy and culture was devastating. These incidents show how the arctic environment and its inhabitants are particularly sensitive to events and influences from other parts of the world.

There are deposits of radioactive minerals of potential economic importance in arctic Canada, Finland, and the Soviet Union. None of these has yet been developed. Because of the vulnerability of arctic ecosystems, and because of international attention to the opposition by local residents and environmentalists to exploitation of the deposits, special environmental assessments and controls, which will take into account international aspects of local decisions, may be needed.

Protection of arctic marine waters from radioactive contamination poses special problems. It has been reported, without verification, that spent radioactive fuel from nuclear engines was disposed of in sub-arctic waters; but it has also been stated that this practice, if it once was carried out, has been discontinued. It is likely that an increasing number—perhaps before long the majority—of vessels plying the Arctic Ocean will be nuclear-powered; although in normal operation these should pose no radioactive hazard to the environment, in the event of accident there is a risk of oceanic contamination by highly radioactive substances in positions where removal or protection will be very difficult. Recent accidents to nuclear-powered submarines in sub-arctic waters have drawn international attention to a problem which could be even more difficult in higher latitudes.

AN ARCTIC POLLUTION PRIMER

Organochlorines

Organochlorines are a large class of chemicals including industrial organics, agricultural organics, by-products of anthropogenic activities, and chemical transformation products. They are important environmental contaminants due to their high stability and persistence in the environment, high bio-accumulation potential, potential for high chronic toxicity, and as a result of the large quantities which have been released into the environment.
Source: Department of Indian Affairs and Northern Development

Polychlorinated biphenyls (PCBs)	Chlordane
Hexachlorobenzenes (HCBs)	Toxaphene
	Endrin
	Aldrin
Hexachlorocyclohexane (HCH)	Dieldrin
DDT	Lindane
DDE	Dioxins
Mirex	Furans

1) Industrial Organics

Sources: Incineration, volatilization

PCBs (polychlorinated biphenyls)
- human symptoms include diseases of blood, immune and nervous systems, respiratory impairment, dermal toxicity, and mutagenic and carcinogenic effects
- linked to premature births, reduced birth weight, smaller head circumference, compromised neuromuscular development

HCBs (hexachlorobenzene)
- porphyria, still births in humans
- developmental abnormalities in rats

2) Agricultural Organics

Sources: spray drift, evaporation, volatilization, wind erosion

HCH (hexachlorocyclohexane)
- banned in 28 countries and severely restricted in five
- spread via evaporation and air transport
- repeated exposure linked to damage of kidneys, pancreas, and nasal mucous membranes
- accumulates in breast milk and fatty tissues

Lindane
- banned in eight countries and severely restricted in five
- affects central nervous system
- liver and kidney damage in test animals
- probable carcinogen
- linked to aplastic anemia
- harmful to developing embryos in test animals

DDT (dichlorodiphenyltrichloroethane)
- causes numbness, parathesia, weakness or paralysis
- carcinogenic

DDE
- causes egg shell thinning in avian species
- reduces egg viability in fish

Chlordane
- linked to cancer in laboratory animals

Toxaphene
- carcinogenic
- extreme toxicity in fish populations (used as piscicide)

Endrin
- banned in 38 countries and severely restricted in 12
- classified as "acutely toxic" by World Health Organization
- has caused cancer in test animals
- prolonged exposure can lead to severe convulsions and, in some cases, death

Aldrin
- banned in 34 countries and severely restricted in 13
- probable carcinogen
- extremely persistent, bioaccumulative
- breaks down to dieldrin in living systems
- believed toxic to fish, freshwater invertebrates, and birds

3) Anthropogenic By-Products

Dioxins/Purans
- possible carcinogens, though still under study
- believed toxic to fish

4) Chemical Transformation Products

Dieldrin
- banned in 34 countries and severely restricted in 21
- used for control of termites, desert locusts, mosquitos, and tsetse flies in developing countries
- extremely carcinogenic
- linked to liver cancer, muscle spasms, tremors, developmental abnormalities
- concentrates in breast milk

Heavy Metals

Heavy metals are metals of high specific gravity. Metals are required for the life processes of animals, including man. The body requires substantial amounts of some metals (for example, sodium and iron) and has a high tolerance for their absorption. Other metals, such as "micro-nutrients" copper and zinc, are required in much smaller amounts, and the body has a lower tolerance. Finally, there are metals which are not essential to life processes and are toxic to both man and other species at lower levels.

The effects of such substances vary with the metal. In some cases, heavy metals attack nervous tissues by enzymatic blocking of biochemical reactions. More frequently, they destroy excretory organs, such as the liver and kidneys.

Beryllium	Mercury	Thallium
Boron	Nickel	Titanium
Cadmium	Selenium	Vanadium
Lead		Zinc

Arsenic
- used in herbicides and defoliants
- found in foods associated with cotton-seed by-products
- linked to weight loss, hair loss, skin lesions, and cancer

Cadmium
- appears as a contaminant in use of zinc
- causes severe gastroenteritis (linked to *itai-itai* disease in Japan)
- associated with chronic lesions, testicular necrosis, fetal malformation, cardiovascular hypertension

Selenium
- suspected carcinogen

Lead
- originates from lead smelters, solder, gasoline
- attacks nervous system
- linked to diseases of kidney and haematopoietic system
- may lead to acute encephalopathy, anemia, nephrosis

Mercury
- organic mercury, particularly methylmercury, highly toxic
- causes severe gastroenteritis, nephritis
- affects brain, causing serious and irreversible damage
- common in aquatic system

Doing the Great Slave samba

What with global warming, tropical forests and the predictable rhetoric of George Bush dominating attention at Rio de Janeiro's Earth Extravaganza, how could you blame northerners for feeling a little left out?

by TED BURACAS

In our own backyard

It isn't as if there are no environmental problems here; the residents of Canada's North, from Yellowknife to the smallest hamlet, know they are not immune. Get beyond the clichés, and you hear complaints about traffic congestion and smog, abandoned scrap yards and rusting fuel barrels. These are local manifestations of the larger problems that affect the entire planet. What is different in the North is the enduring link with the land. When the land is harmed, the people are harmed. And they don't need reams of reports and studies to tell them so.

Yet, environmentally-speaking, the Arctic was "discovered" only in the last decade. Now, of course, it has become trite to point out that global perturbations are just that—changes that affect every region of the planet. The tired clichés—"undisturbed, pristine, isolated"—are no longer in vogue, the vision of a boreal Arcadia they evoked revealed as a clever fiction.

But while the long-range transport of pollutants is an increasing problem, so are local sources. Over the past fifty years, lax government regulations and ambivalence toward the environmental costs of development have created local hot spots throughout the territories. The battle between jobs and environmental protection continues today, fueled by continuing economic recession.

A case in point is the mining industry, a mainstay of the northern economy, but one which has had a significant long-term impact on the northern environment and the people who rely on it. While there are now only six producing mines in the Northwest Territories, dozens of abandoned sites, open pits and tailings ponds continue to pose an environmental hazard. Pine Point, a lead-zinc mine operated by Cominco from 1965 to 1989, is just one example. Don Balsallie, a member of the Deninoo (Fort Resolution) Community Council, recalls how his father and uncles were ingloriously evicted from their traplines and cabins by bulldozers.

"They just came in with Cats and said, 'You can't be here anymore,'" Balsallie recalls. "Their traplines were taken away and then they had to live with the chemicals, the dying trees, and the animals that were scared away. I remember one elder who recently passed away; he was still waiting for an apology."

When the ore gave out, the town folded, leaving deserted buildings, eighty-two open pits and tailings ponds. Some of the ponds eventually dried up, leaving the dust—which Balsallie and others contend is toxic—to blow into the surrounding area, including Great

Courtesy of Canadian Arctic Resources Committee

Slave Lake. The huge open pits remain, a legacy to short-sighted development, at least in the eyes of local residents who must live with the effects.

The Pine Point experience is not a rarity in the Northwest Territories. But Cominco has at least acknowledged responsibility for the site and is working, however slowly, to stabilize it. For the other abandoned mines that desperately need attention, there is little hope other than for government money to pay the bill. To Tom Heofer, executive director of the Northwest Territories Chamber of Mines, this is not an unreasonable expectation. "Companies should not be expected to go back ten or twenty years after they have finished with a site and bring those sites up to today's standards," Hoefer says. "The rules have changed since then. Today we have rehabilitation bonds to ensure that mines clean up after themselves. Back then, they played by government rules that were not as strict."

The federal government was to have restored the sites with money from the Arctic Environmental Strategy (AES), part of the multi-million-dollar Green Plan. Trouble is, they don't know how to do it. Engineering reports on three of the sites most in need of attention—Rayrock and Discovery, just north of Yellowknife, and the Rankin Inlet nickel mine—are due soon, and the price tag could well exceed $5 million.

No one who lives in the North can doubt the importance of the mining industry to the economy. It is the largest private-sector employer in the Northwest Territories, employing between 1,700 and 2,000 people. Production value for 1991 exceeded $500 million. In recent years, even long-time opponents have applauded the industry's willingness to change its heavy-handed ways, but the change has been slow in coming. And many critics say there is still a long way to go.

Like mining, oil and gas production is an integral part of the northern economy. Companies have spent billions seeking, extracting, and transporting oil and gas to thirsty southern markets. The pace of exploration has slowed in recent years, but the pressure to develop remains.

During the summer of 1990, Gulf Canada Resources drew heavy criticism from the Inuvialuit Environmental Impact Review Board for shortcomings in its exploration plan. A major player in the offshore, Gulf eventually scuttled its Beaufort drilling schedule for the year after a squabble broke out between the review board and the Canada Oil and Gas Lands Administration (COGLA) over jurisdiction. COGLA, for its effort, has since been disbanded.

Today, in the communities of the lower Mackenzie, the mood is mixed. Many communities fully support such development because of the employment it offers. However, it may all be a moot point; Gulf has undergone a major restructuring and disbanded its entire northern operation, eliminating for now its participation in the region. The company blames the economic downturn and a weak market for crude oil, but the uncomfortable reception it received two summers ago could not have helped either.

INUIT CIRCUMPOLAR CONFERENCE 1992

INUVIK AND TUKTOYAKTUK
JULY 20 - 24, 1992

TOWARD A MORE SUSTAINABLE FUTURE

Keynote Dinner Address
Monday, July 20, 1992

J.M. MacLeod
President & CEO
Shell Canada Limited
Calgary, Alberta

Chairman, distinguished head table guests and distinguished delegates.

It is a great pleasure for me to participate with you in this Inuit Circumpolar Conference and it is an added pleasure and honor to be able to speak to you this evening.

My company, Shell Canada, hopes to have the opportunity some day to engage in circumpolar development – right here in the MacKenzie Delta. But first we must follow the crude oil discovery we made in 1990 with the discovery of sufficiently large volumes of oil to provide the economic impetus for a sustainable development project. We plan to conduct further exploration over the next few years to determine whether or not those large volumes of oil can be found.

You will notice that I said a sustainable development project – not an oil development project. The resource that we hope will provide the development opportunity in this case is indeed oil. But whether or not such an opportunity will be realized depends on two things. First, will sufficient oil be discovered and second, if discovered, will development of the oil be sustainable in the context of economic, environmental, social and cultural considerations.

Courtesy of J. M. MacLeod

When invited to speak to you this evening I responded that I would be happy to do so if my topic could be sustainable development. Later I was delighted to learn that the purpose of the Conference, in the words of your organizers, is to..."address economic, social, cultural and environmental issues important to the Inuit living in the Circumpolar region." I was delighted because if the Conference does address economic, social, cultural and environmental issues important to the Inuit in the region, I suggest that in fact it will be dealing with sustainable development.

My frame of reference for Sustainable Development is the definition given the term by the Bruntland Commission that described Sustainable Development as "...development that meets the needs of the present without compromising the ability of future generations to meet their needs."

This definition does not give much guidance to plans and actions that would lead to a more sustainable future however, without some operational elaboration. A somewhat more specific statement that I have worked with as an operational framework for some two years now, goes beyond the simple Bruntland definition, to acknowledge that achievement of a more sustainable future will require us all to integrate economic, environmental, social and cultural issues in decision making.

That, of course, is the heart of your statement of the purpose of this Conference.

As each of us proceeds to play some role in implementing sustainable development it is well to recognize that there are large bodies of opinion throughout our global society, starting here at home, that are cynical, skeptical or suspicious of the value of this concept that is still new to Western scientific thinking, although an inherent part of the traditional knowledge of aboriginal peoples. There are bodies of opinion that reject the concept.

I have become committed to the concept personally, and in my role as leader of my company. You might well ask why and how.

Part of the answer to that question arises from the fact that I have had senior executive responsibility for various large units of Shell Canada's operations for two decades, including the responsibility for environmental management related to those operations. Now my responsibility encompasses all the company's activities. At times during those years I have observed directly the confrontational approach to environmental management between industry, environmentalists and, at times, governments.

These experiences have given me a sensitivity to the importance of environmental stewardship and, at the same time, a concern about failures of the adversarial, confrontational process to achieve committed, effective effort.

More recently, I have been fortunate to have the opportunity to play an active role in pioneering sustainable development activity both within and outside the company.

Within Shell, my senior management colleagues and I have led effort that will result within about a year in the application of sustainable development principles to all of Shell's activities. We have built this effort upon an explicitly stated Corporate Sustainable Development Policy.

External to Shell, I have been a member of the National Round Table on The Environment and the Economy since its inception in 1989. Round Table effort has led, in turn, to an initiative by a multipartite industry, government and NGO group of people that I chair, to stimulate the integration of sustainable development material into the formal curriculum for kindergarten to Grade 12 in all provinces and territories in Canada. You may be interested to know that the color of our program brochure is Arctic Blue.

I have also been a member and Co-Chair of a Sustainable Development Task Force within the principle Association of business leaders in Canada. Our Task Force has produced a statement of Sustainable Development Principles that we have recommended be applied in all the activities of the 150 members of the association.

The fundamental objective of each of these initiatives has been to help move our society toward a more sustainable future through influencing change in individual, corporate, institutional and government behavior. Most importantly, constructive results are being achieved in each of the initiatives through the learning experience of consensus building between participants of tremendously diverse backgrounds but who share a dedication to work, each in their own sphere of interest and influence, toward relieving the stresses of development on the global environment.

The threat to the lives of future generations that might result from emission of greenhouse gases, from ground level ozone and from depletion of stratospheric ozone has most certainly elevated the level of concern for the environment in the global community. That concern is a key driver of sustainable development.

It is not clear to me however, whether climate change will in fact become life-threatening. The hypotheses and evidence lend themselves to conflicting conclusions. But I do not believe that one must be in terror of global warming to become committed to sustainable development.

Consider a point made repeatedly by Jim MacNeill, who is a colleague member of The National Round Table on Environment and Economy, a former member, and Secretary General, of the World Commission on Environment and Development and principle author of Our Common Future (the "Bruntland Commission" report). The point MacNeill and others make is – if the world population follows present trends to double to 10 billion people over the next forty years, the planet would be pushed beyond some critical limits to produce the fourfold increase in food calories, the sixfold increase in useable energy and the eightfold increase in income needed to satisfy the population, applying today's development standards.

Quite aside from whether we are on the way to serious climate change, the implications of those simple statistics are a powerful motivator towards the reforms that sustainable development concepts are intended to achieve in the manner in which we develop and use global resources in future – including environmental resources.

I leave with you the proposition that there are multiple motivational influences on us as individuals, and in our roles as leaders of communities, corporations, institutions or governments, to make what contribution is reasonably within our power toward a more sustainable future.

I have been speaking very personally about my own commitment to sustainable development and the influences motivating all of us as individuals to the same commitment.

I want to go beyond the individual now and say a word about the corporation.

Corporations have had mixed records in environmental management over the past two decades. Some records are good, some otherwise good records have been marred by environmental accidents, and some records have been poor, reflecting insensitivity to sound environmental management practice. I believe it is clear that in the future the corporation has no choice. Because all stakeholders in the corporation are now intensely concerned about the environment, beginning with employees, the corporation must ensure that all of its operations are environmentally sustainable or suffer loss of its legitimacy – in effect loss of its mandate, from employees, customers, other stakeholders and the public.

The Sustainable Development Policy we created in Shell Canada in 1990 was driven by the personal motivations of members of our senior management and our collective determination to protect the legitimacy of our corporation. It sets forth clearly the commitment of the corporation to sustainable forms of development. We state in the policy that:

- we will apply Sustainable Development principles to all Shell activities;

- we will implement Sustainable Development self monitoring mechanisms;

- we will evaluate public opinion on Sustainable Development; and

- we will participate in consultative processes on Sustainable Development.

There are no reservations, contingencies or exceptions to that commitment.

Perhaps of particular interest to the Inuit people from northern Canada participating in this Conference, there will be no exception for developments affecting aboriginal peoples. In fact I believe the multipartite consensus building processes that are essential to achievement of circumpolar sustainable development will provide an exceptional opportunity for us all. Any opportunity to learn the benefit of applying the traditional knowledge of aboriginal peoples in harmony with western scientific knowledge in sustainable development decision making.

Thank you all for your attention this evening. I hope you sense from my comments my personal commitment and that of my corporation to sustainable development. Of far greater consequence, I hope you sense as I do, that sustainable development is a concept that, without doubt, society is compelled to adopt.

Best wishes for the success of your conference.

Nuclear waste: a time bomb for the Arctic?

COVER STORY

A shadow over Siberia

Soviets' nuclear waste eyed as latest threat to Arctic environments

BY MIRO CERNETIG
The Globe and Mail
NOVOYE CHAPLINO, Russia

In the icy grip of winter, Siberia can still seem primordial — even pristine. Jagged mountain peaks rise into an indigo sky; the blinding white snowscape stretches into the empty curve of a frozen ocean.

The illusion of environmental purity doesn't last for long, though, when you tour Siberia's easternmost reaches with Vladimir Etylen, the president of the little-known Russian republic called Chukotka.

As the military half-track plows its way out of a deep valley, its steel treads churning up the snowdrifts, Mr. Etylen's nose crinkles with consternation and he points out the vehicle's window.

A black cloud looms in the distance, belched onto the horizon by a coal-fired power plant in Novoye Chaplino, a village on the frozen coast of the Bering Sea.

"Look," Mr. Etylen urges, yelling above the guttural roar of the diesel engine. "Pollution — here too."

With the end of the Cold War that once kept Siberia closed to outsiders, Mr. Etylen now freely points out the legacy of unbridled military and industrial development: choking air pollution, heavy-metal contamination on the tundra, rivers and bays laced with toxic chemicals and devoid of fish.

And he is at liberty to talk about the most chilling inheritance of all: nuclear waste on the ocean floor.

For decades the Soviet military secretly dumped used nuclear reactors and radioactive waste off Siberia's Arctic and Pacific coastlines.

The material now sits on the shallow ocean floor, where it may pose a threat not only to Russia but also to Canada and other countries sharing the delicate Arctic environment.

"We did not know this was happening," Mr. Etylen says. "Now we are seeking international advice and assistance on this matter. For help from others."

No one can say for sure what danger the nuclear dumps pose. "The big question is how long is it going to take until the sites start to decay, or corrode, and enter the environment?" says Michael Bewers, a marine specialist with Canada's Bedford Institute of Oceanography and chairman of the international scientific committee investigating the hazard.

Past experience shows, however, that Siberia's problem should be every circumpolar country's concern. For the delicate Arctic environment magnifies the effect of all forms of pollution on the people who live at the top of the World — including Inuit, Indians and Laplanders.

Unlike southerners, who eat processed foods that are monitored for contaminants, the natives of the North eat foods that are not: caribou, whales, polar bear and other animals.

These are creatures at the top of the food chain which pick up trace levels of toxic pollutants: polychlorinated biphenyls, DDT and heavy metals such as cadmium and mercury that come with industrialization.

Invariably, these pollutants end up in meat-eating humans, posing concerns to researchers in virtually every Arctic country about the long-term health effects.

The same thing occurs with radioactive particles, known as radionuclides, which can cause cancer and genetic defects. This was proved decades ago when the United States and the Soviet Union tested nuclear bombs in the atmosphere. Radionuclides travelled up the food chain from Arctic lichen to caribou to people.

Even today, when scientists say the amount of radionuclides has dropped to safe levels, some northern residents continue to feel a bit like guinea pigs.

"I got myself tested last year," says Charlie Johnson, an Inuk from Nome, Alaska, who sits on the U.S. Arctic Research Commission. "They could tell I ate lots of sea mammals because of the high levels of heavy metals in my body. They could tell that a Canadian [Inuk] they also tested was eating lots of caribou in her diet, because she had more radionuclides in her body than I did."

Industrial pollutants are still blowing into the Arctic from faraway sites

Courtesy of The Globe and Mail, March 6, 1993

in North America, Europe and Asia.

And the Soviet undersea nuclear dumps are fuelling renewed concern about radioactive contamination, according to Garth Bangay, who heads Canada's $100-million Arctic Environmental Strategy — a plan to clean up the North and investigate contaminants.

There is some encouraging news. Since the dumps are underwater, Siberian radionuclides will infiltrate the environment more slowly than they would if they were released into the air. But they will eventually leak out, at a rate that depends on their type and location, and on the effect of undersea currents.

Such data are unavailable to anyone outside Russia, and to almost everyone inside it.

Some of the answers should come this spring, when Russian President Boris Yeltsin is expected to release a detailed report on Arctic nuclear waste.

In the meantime, the best guess comes from Charles Hollister, a senior marine geologist at the Woods Hole Oceanographic Institute in Massachusetts, who studies the issue for the U.S. State Department and has interviewed Russian officials.

He estimates that the Soviet Union dumped at least 13 fuel-containing reactors onto the Arctic continental shelf, together with nine whose fuel probably was removed and about 11,000 canisters of assorted nuclear waste.

"The total radioactivity involved is about the equivalent of two Chernobyls," Dr. Hollister says, referring to the Russian nuclear power plant that propelled radionuclides around the globe in 1986. "Put another way, that's about the equivalent of half of all the radioactivity released during atmospheric nuclear testing during the Cold War."

Most material is in the Barents and Kara Seas, around the island Novaya Zemlya, where the Soviet Union carried out tests of nuclear bombs.

But some is believed to have been dumped in the Northwest Pacific off the Kamchatka Peninsula, where there are also nuclear submarine bases.

Dr. Hollister says it is far too early to panic. Nuclear dumping occurred between the 1960s and 1980s, and there is no sign so far that the Arctic Ocean is any more radioactive than any other. That could mean the dump sites are not leaking significantly, or it could mean leaked radioactive material is not spreading.

Scientists who met four weeks ago in Norway under the auspices of the International Atomic Energy Agency agreed there is no evidence so far of environmental damage to the Arctic Ocean from Siberia's nuclear dumps.

But those who study the question are hoping Mr. Yeltsin's report will answer four key questions, according to Hugh Livingston, a marine radiochemist at Woods Hole.

• Were castoff reactors sealed strongly enough to ensure no leakage for 500 years, as some Russian specialists have claimed?

• Has Russia ever tested for leakage around the sites?

• What kind of radioactive material was dumped? Some radionuclides break down quickly while others, such as plutonium, last for tens of thousands of years.

• Are there maps showing the undersea dump sites, to help calculate the effect of ocean currents?

Good maps are especially important, according to Dr. Livingston. It will be essential to explore the waste sites — but it could be hazardous, not because of radiation but because old weaponry may have been dumped in the same area. "If you're not careful, you could get blown up just going in to take a look," Dr. Livingston says.

It spurred the international Arctic scientific community to redouble its efforts to co-ordinate research and communicate findings. The eight nations are now sharing research in several key areas: persistent organic contaminants such as PCBs, DDT, and toxaphene; oil pollution; heavy-metal contamination; noise pollution, which can disturb wildlife migration; acid rain and radioactive contamination.

Until recently, radioactivity was considered a low priority. Studies on the tundra, many of them carried out in Canada, found radionuclide contamination had dropped since the days of atmospheric tests. Even the Chernobyl disaster did little to increase radionuclide levels in most of the Arctic.

But now Canadian scientists are seeking financing for seven projects geared to measuring radioactive pollution. They want to measure radionuclide levels in mussels taken from Arctic waters, as well as in the inner core of the ice cap, where they hope to gain historical data.

Similar research is under way in Norway. The United States is spending $10-million on studies aimed at limiting environmental damage from the underwater dumps.

A Norwegian-Russian research mission in the Barents Sea last summer took sediment and water samples, some of which are being analyzed in Canada. The results will probably be released in August at a conference in Kirkenes, Norway, and will give the first indication of whether there is significant nuclear leakage.

Even if there is no evidence of major leaks now, the dumps will eventually pose a problem as corrosion takes its toll. So attention is also being focused on the type of material dumped.

Siberia's "atom problem" as some Russians call it, has yet to filter down to the level of the average citizen in Chukotka, which is nine time zones away from Moscow. Asked about the nuclear dumping, most people simply shrugged.

"Our press does not write about this subject much," explained Mr. Etylen, who like most Russian leaders — and journalists — spends most of his energy worrying about the economy.

But word of Siberia's industrial and nuclear nightmares is getting around, thanks to the easing of travel restrictions inside the former Soviet Union and abroad. And the region that was once viewed as little more than a place to plunder raw resources and park political prisoners is now home to a heightened environmental consciousness.

Mr. Etylen, one of 16,000 Chukchi natives in the region, tells visitors he wants "development with Western environmental standards" in Chukotka, a republic of 200,000.

In Novoye Chaplino — home of Russia's remaining 1,200 Inuit, known as the Yupik — a handwritten poster taped to a schoolroom wall warns in bold black letters: "....Abundant wildlife and bountiful habitats have survived because so much of the region is underpopulated. However, pressures of a rapidly developing world are leaving warning signs."

Siberians, in fact, are eager to take part in international co-operation on the Arctic environment — which led to the creation in 1991 of an Arctic Environmental Protection Strategy. It was signed by all eight circumpolar countries: Canada, The United States, Russia, Norway, Finland, Sweden, Iceland and Denmark.

They agreed to measure the extent of Arctic pollution and eliminate its sources, using international treaties where possible. More significantly, they are pooling research resources. The latest example is an atmospheric testing station Canada is helping to build on an island in eastern Siberia that will measure air contaminants from Russian industry.

The importance of such co-ordination became clear in 1985, when Canada's department of Indian and Northern Affairs released the first comprehensive compilation of Arctic pollution research. Known as the Wong Report after its author, Michael Wong, the report shattered any illusion that the polar region might be immune to industrial pollution simply because of its remoteness.

If it is likely to produce radionuclides such as plutonium 239 or plutonium 240, a Canadian government report predicted, contamination will probably be confined to small areas. Such isotopes last for thousands of years but tend to attach themselves to ocean-floor sediments.

The report recalled a U.S. B-52 bomber that crashed into the ocean off Thule, Greenland, in 1968, spilling radioactive plutonium. After 25 years, it said, little of the material had travelled more than 50 kilometres from the crash site.

It would be more troubling if radionuclides such as cesium 137, strontium 90 and technetium 99 were leaking. These break down faster than plutonium but are easily transported in water and "ultimately would be widely dispersed throughout the Arctic Ocean," the Canadian report concludes.

There are several options for dealing with the dump sites. Dr. Hollister says that if released radionuclides were quickly diluted, or attached themselves to surrounding sediments, a slowly leaking site could simply be left alone and monitored. In some cases the sites could be capped, other scientists say.

Recovering the material is considered risky and expensive. It also raises the question of what to do with the stuff.

Some scientists believe the best way to dispose of highly radioactive material is to turn it into a hard glass-like substance and bury it in sediment at the bottom of deep ocean basins. "If I had my choice of where to put all this stuff, it would be on the deep-sea floor," Dr. Hollister says.

Some environmentalists disagree, saying too little is known about currents and subocean seismic activity.

Although scientists are clearly dealing with the most troubling case of nuclear mismanagement since Chernobyl, Dr. Hollister says there is some cause for thanks.

"Things could be worse," he says. The Russian military is still believed to have more than 100 nuclear-powered submarines and surface vessels. Each has two nuclear reactors. If the Cold War had continued, many of them would probably have been destined for shallow, secret graves in the Barents Sea.

Questions and Concerns

READING 6

1. Entire populations of an Arctic species like the Peary Caribou can die if normal climatic conditions change. Why are they so fragile?

STORMY WEATHER

AS CLIMATIC CONDITIONS CHANGE, SO DO THE FORTUNES OF THE PEARY CARIBOU

By Ed Struzik

WALKER RIVER, BATHURST ISLAND — In the first three weeks of June, Canadian Wildlife service scientist Frank Miller had flown about 30 hours of helicopter time searching for the tiny Peary caribou on this island in the High Arctic. Seldom, however, did he see more than 50 animals in a day, even though he had estimated there to be about 800 on the island.

At Miller's Walker River base camp, the possibility of a catastrophic die-off or mass migration to another island to the south was in the back of some people's minds. But Miller wasn't buying any of it.

"If there was a catastrophic die-off, we'd be seeing carcasses by now," he said as we added more hours of flight time in mid-June. "But that just hasn't happened. We've yet to see a single dead animal. And there's no evidence to suggest that these caribou are moving anywhere but maybe to the smaller islands to the west or to the north of Bathurst Island where we haven't been able to get to yet because of the bad weather. I'm convinced that these animals are doing exactly what they're supposed to be doing. They'll turn up."

After another six hours of flying, and a near complete circumnavigation of Bathurst Island, there were again only a few caribou to be seen. Miller, I suspected, must have been having doubts. But then on the last leg of the flight, a group of nine animals appeared on the horizon, then another group of six, and one of seven. Within 20 minutes, we had counted 71 heading up the east coast, just as Miller had been predicting all along.

"You can cover a lot of territory, and miss a lot of caribou along the way," he said when we finally touched down. "Sometimes, it's a case of just not seeing them in between the snow patches in this dull light. Other times, they can be inland, or behind a hill, and you'll fly right by them. In July, when the weather is clear and the snow is gone, they'll be much easier to spot."

Courtesy of Ed Struzik

IF ANYONE SHOULD KNOW WHAT IS HAPPENING TO PEARY caribou in the High Arctic, it's Miller, a bullish American ex-marine whose gravelly voice and volatile temper is forever at odds with a wicked sense of humour and a soft spot for caribou.

"To paraphrase another biologist who worked in the North in the 1950s, no other animal exhibits so close an approach to a Garden of Eden trustfulness of man as the caribou," he says. "So in my mind, no other is more worthy of being cherished and safeguarded in their natural haunts. Working with them for 26 years has afforded me some of the most marvellous sights in nature. It has reinforced in me the belief that the Arctic would indeed be an empty land without them."

With well over a million caribou roaming over Canada's mainland tundra, there is little likelihood of that happening in the near future. But the situation is just the opposite in the High Arctic where the Peary caribou, a distinct subspecies that occurs only in Canada, make their living. Since 1961, when biologist John Tener flew over the Queen Elizabeth Islands and estimated there to be nearly 26,000 animals, the population has crashed to the point where there are only about 3500 to 4000 left today. And with the exception of Bathurst Island, those numbers are continuing to decline.

"We are rapidly getting to the point where the Peary caribou population will not be able to rebound for several decades," says Miller. "A worst-case scenario is a die-out on most or all of the Queen Elizabeth Islands where nearly all the Peary caribou live. The situation, whichever way you look at it, is very critical."

The Committee on the Status of Endangered Wildlife in Canada (COSEWIC) thinks so as well. It is the official organization of government and private sector scientists who determine whether or not a species of wildlife requires legal protection. After examining data that Miller presented to them at their annual meeting this year, they uplisted the Peary caribou from the serious classification of "threatened" to the more dire level of "endangered." The designation is the most extreme, and means that the animal is seriously at risk of becoming extinct unless man intervenes in some way to manage it back to healthy numbers.

Whether man can do much, however, is questionable where the Peary caribou is concerned; for climatic factors, rather than loss of habitat, overhunting, or other direct actions by man, are really at the root of its demise. In fact, the only people to hunt Peary caribou — the Inuit of Resolute on Cornwallis Island and Grise Fiord on Ellesmere Island — voluntarily stopped shortly after being told of the decline.

Ironically, it is Miller, the landed immigrant from the United States, who is one of the few non-natives to see the Peary caribou's possible extinction as a tremendous loss for this country. "It is the only large animal that occurs exclusively in Canada," he says. "And as far as we know, it has been here for thousands of years, its ancestors probably having survived the last ice age in the High Arctic. No other large-hoofed animal other than the muskox has adapted as well to such extreme conditions, at least up until now."

IN RETROSPECT, THERE WAS NOTHING TO WARN SCIENTISTS that the Peary caribou was in trouble in the years that followed Tener's survey. In fact, when Miller first got involved in studying them back in 1972, it was because most everyone assumed that the population was increasing, and that more animals might be allotted for the annual native harvest and for the introduction of sports hunting. Essentially, Miller's orders were to see how many more could be taken without hurting the population.

In that first survey year, however, Miller counted only about 1400 animals; there were no calves among them, and very few yearlings. Although severe weather had prevented him from getting to all of the islands he had targeted, it was clear that something terrible had happened. Exactly what it was, the scientist couldn't figure out.

"I was nervous," he recalls. "It was my first time out on this survey and I thought 'My God, I must be screwing up.' But I was pretty sure that I couldn't be missing that many animals. And the fact that I had seen no calves and very few yearlings only reinforced that belief."

The following year produced better weather and a more representative survey of the islands. And this time there were plenty of claves. But again, there were very few caribou to be found anywhere. Miller estimated there to be 5,143 animals, nearly 75 percent less than what Tener had counted just a decade earlier.

WITH TWO YEARS OF DATA AND A PRETTY CLEAR PICTURE OF the population, all that really remained was to discover what was killing them off.

It didn't take long for Miller to find out. The next year not only failed to produce more calves, there were very few yearlings as well, meaning the last year's newborn had died. In addition, Miller found a large number of carcasses all throughout the southwestern portion of the Queen Elizabeth Islands where 80 percent of the animals dwell. By the time the counting of the dead and living was completed, he estimated that nearly half the animals that had been there in 1973 were dead in 1974.

"That winter had been a particularly severe one," explains Miller. "Such were the conditions that ice formed on the ground in autumn. It was followed by a heavier than normal snowfall in the winter and a spring thaw that started early, then stopped with below-freezing temperatures that persisted through to the early summer months. Basically, the caribou just couldn't crater through the snow cover and ice to get to the vegetation that they normally use to get them through to the summer months."

Illustrative of the Peary caribou's plight are some snow-ice transects that Miller has set up around his camp. The longest is 7.5 kilometres, starting from the sea coast and extending inland to the highest point of land. In May when the camp starts up, Miller and his assistants track snow depths until the melt starts. Then he begins checking for the ground-fast ice. This occurs when the melting snow percolates down into the colder ground. Because this ground is often -10°C to -20°C, the warmer melting snow will freeze into a layer of ice below. This year the deepest layer of ground-fast ice was 15 centimetres. In past years, Miller has measured up to 30 centimetres of ice.

"It takes a lot of effort to chop through it with an axe," says the scientist. "So you can imagine a caribou trying to get through the same thing. If this ice is as all encompassing and thick as it probably was in the winter of 1973-74, then the caribou have no chance. They really are dependent on there being at least some open ground during that time."

THE PEARY CARIBOU'S UNFORTUNATE PLIGHT, HOWEVER, didn't end in 1974. Miller's Canadian Wildlife Service colleague, Don Thomas, later discovered it took three years for the female caribou to return to a normal reproductive rate after that disasterous winter. In fact, during the first two years following the major die-off, less than 7 percent successfully gave birth to calves.

"There's evidence of major die-offs among other ungulate populations," says Miller. "But for there to be virtually no reproduction for five successive years is unheard of."

Whether Peary caribou will be subjected to similar freezing episodes and subsequent catastrophic die-offs in the future is difficult to say. But according to Miller, global warming brings with it the potential for a worst-case scenario. "In the long run, a warming trend is not likely going to hurt the Peary caribou," he says. "It's the transition to that warming that is the real concern. If it brings with it more fluctuations in weather and temperatures that produce the ground conditions we saw in 1973-74, then the caribou aren't going to stand much of a chance, especially if they're at the very low numbers that they are right now."

In the coming years Miller plans to continue his annual surveys, but confine them to Bathurst Island. There's just not enough money available to allow him to fly over other important sites on Melville and Prince Patrick islands further to the west.

It's a fact of life that Miller has resigned himself to in this era of cutbacks. What he would like to have, though, is 10 satellite transmitters which would allow him to follow the movement of caribou year-round. The equipment would allow him to track the animals on Bathurst and perhaps other sites on the Queen Elizabeth Islands in a way that would be impossible or prohibitively expensive otherwise.

"If we really knew how they used their space seasonally and on a daily basis, it would give us a better appreciation for their needs and their important habitat," he says. "With that information, we could determine what areas deserve special protection in the event that oil and gas development or mining decides that the High Arctic islands are valuable to them. The bottom line right now is that we don't know how the animals use their range to make these kinds of biologically sound predictions."

Reading 7

Questions and Concerns

READING 7

1. The Northwest Territories was the first jurisdiction in Canada to allow individual citizens to take legal action on behalf of the environment without having to demonstrate personal damages. Why do you think the NWT government felt this law was necessary?

Citizens for the environment: is the sword too blunt?

LAST YEAR, the Northwest Territories became the first jurisdiction in Canada to legislate the right of individuals to take action on behalf of the environment. The Northwest Territories Environmental Rights Act was introduced as a private member's bill by MLA for Yellowknife Centre Brian Lewis in 1990. Among other provisions, the Act allows any two individuals to request information or action by the Government of the N.W.T. on behalf of the public. The government, in return, is obliged to respond within sixty days. But that doesn't guarantee much will happen, as Yellowknife resident Kevin O'Reilly found out.

"We wanted the Minister [of Renewable Resources] to investigate the stack emissions from the Giant Yellowknife Mine," says O'Reilly, who is a director of the Yellowknife-based environmental group Ecology North. "What we got at the end of sixty days was a letter asking us what evidence we had to think Giant posed an environmental risk to begin with." Needless to say, they had hoped for better.

The success of the legislation rests on the premise that the average citizen has an interest in the environment and is willing to pursue it. Environmentalists say it's a way to make governments more accountable when it comes to environmental protection. And government officials admit that public interventions give them the ammunition they need to get action on issues previously ignored.

Empowering individuals to take action on behalf of the environment may not be a new concept, but northerners are still trying to figure it all out. There have been only two interventions filed since the Act became law.

Courtesy of Canadian Arctic Resources Committee

Questions and Concerns

READINGS 8 & 9

1. The destruction of the world's ozone layer is occurring more rapidly than scientists predicted. What does this dilemma say about the vulnerability of the Earth and the way human beings have allowed economic affairs to take precedence over the security of the environment?

Sizing up the strategy

LAUNCHED WITH MUCH FANFARE a year ago, the $100-million Arctic Environmental Strategy (AES) is the northern component of the Government of Canada's Green Plan. Intended as a practical response to northerners' concerns that the arctic environment was being ignored, the AES set out to identify chemical contaminants, clean up waste sites, upgrade water monitoring and pollution control, and help integrate environmental and economic concerns.

The AES is actually composed of four separate action plans: action on contaminants, action on water quality and quantity, action on integrating environmental and economic concerns, and action on waste management. Although the money comes from Ottawa, local officials have a surprising amount of autonomy and will be able to sign off about three-quarters of the funds in the North. But the two territories are running their own shows, and this fragmentation causes problems, environmentalists say. Without the cohesion of governments and environmentalists working together, they say the strategy amounts to little more than a patchwork of clean-up projects lacking a long-term vision.

Floyd Adlem of the Department of Indian Affairs and Northern Development (DIAND), who heads up the N.W.T.'s portion of the $30-million action plan on waste management, has a daunting task ahead of him. His department has identified the DEW-line clean-up and three major abandoned mine and tailings sites as priorities. DIAND has taken responsibility for the rehabilitation of these sites. In some cases, the original owners cannot be found (their companies having been dissolved), and in others, ownership was transferred back to the Crown from other departments.

Critics charge that DIAND got suckered into accepting responsibility for U.S. military bases abandoned during the early 1960s. Privately, some department staffers agree, saying that the United States, or at least the Department of National Defence, should pay for cleaning them up. Adlem says DIAND fairly accepted those sites under the standards of the day, and that there is no point trying to lay blame now. He admits, though, that cleaning up these sites alone could cost from $100 million to $300 million.

The jewel in the AES crown may well be the relatively small Environmental Action Program initiative, another part of the effort to encourage environment/economy integration. Pegged at $200,000 for both the N.W.T and the Yukon, the program funds community-oriented projects that educate people and benefit the environment. In the Northwest Territories, money has gone to support community recycling operations and numerous school-based environmental education initiatives. The fund has even donated to a study that looks at the utility of using sled dogs rather than snowmobiles.

Lorraine Hewlett, a director of the Yellowknife environmental group Ecology North, says the program has been successful in putting money into the hands of local people. "It deserves lots of applause," she says. "It pulls together so many groups within the community—what we would call environmental partners. And it allows things to happen at a quicker pace than if one group does it on their own."

Though less sanguine about the future of the AES and critical of its underfunding, environmentalists and aboriginal groups have made it clear that they are willing to work alongside the bureaucrats.

"The attitude is very positive," says DIAND's Angus Robertson. "Sure, there are strong criticisms, but there is also strong support from the Dene and Métis nations."

Robertson also agrees that funding is tight. "Yes, we are underfunded," he says. "Any program can use more dollars. It's the old story of whether the glass is half-full or half-empty. What we do have is a godsend."

Courtesy of Canadian Arctic Resources Committee

Underwater noise

Concerns have been raised, particularly by native hunters of wildlife, about the potential negative effect noise may have on wildlife behaviour. Although most of the reactions are not well understood, biologists are concerned that noise from human activities, such as marine traffic, aircraft overflights, and oil and gas drilling on artificial islands, may interfere with echolocation, communication, and the normal behaviour of marine mammals in the immediate vicinity (Finley et al. 1984; Miller and Davis 1984; Richardson et al. 1985).

Although moving sources of noise, notably boats and aircraft, may cause short-term behavioural reactions and temporary displacement of various marine mammals (Finley et al. 1984; Miller and Davis 1984; Richardson et al. 1985), research directed at bowhead whales has not demonstrated a correlation between industrial activities and annual distribution problems (LGL Ltd. 1986). Notwithstanding the lack of direct evidence of serious consequences from noise disturbance, native hunters from Canada and Greenland are concerned that avoidance responses by beluga and perhaps narwhal could affect their movement into traditional harvest areas and thereby decrease hunting success.

Noise disturbance is not only a problem in underwater ecosystems; on land, it has had negative effects, even including deaths. Aircraft overflights of seals and walrus at haul-out sites can cause mortality through stampedes and abandonment of young (External Affairs Canada 1991). Labrador Inuit have been concerned since 1979 about the potential negative effects on caribou of sudden noise caused by low-level military aircraft, and their concern has increased with the increase in flight activity since that time (Rowell 1990).

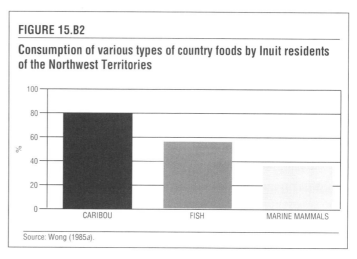

FIGURE 15.B2

Consumption of various types of country foods by Inuit residents of the Northwest Territories

Source: Wong (1985a).

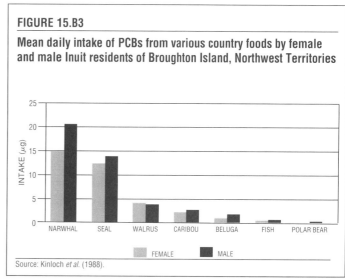

FIGURE 15.B3

Mean daily intake of PCBs from various country foods by female and male Inuit residents of Broughton Island, Northwest Territories

Source: Kinloch et al. (1988).

FIGURE 15.4

Average radioactivity in air and average annual deposition of radioactivity in the Canadian Arctic

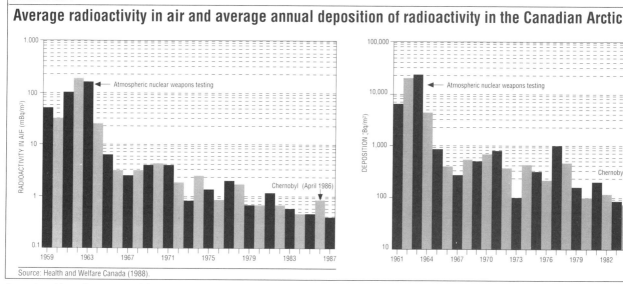

Source: Health and Welfare Canada (1988).

Reproduced from: Government of Canada, 1991, *The State of Canada's Environment*.

TABLE 15.6

Predicted impacts of global warming on the arctic ecosystem

Type of impacts	Predicted impacts
Physical	• virtually all of the terrestrial Arctic is underlain by perennially frozen ground (permafrost), some of which would melt, increasing the active layer, disrupting natural drainage patterns, and releasing methane, thus further enhancing the greenhouse effect • the tree line would gradually shift north, depending on soil conditions, by up to 750 km in the District of Keewatin at a migration rate of 100-250 km per decade, 25 times faster than under normal conditions • many regions would experience significant change in water availability as precipitation patterns change • lakes would experience a longer ice-free season of up to an additional 120 days in the fall and 15 days in the spring • thermal expansion of the oceans and melting of glacial ice could elevate mean sea level 0.5 m or more, causing beach erosion and flooding by storm surges, particularly on the Beaufort Sea coast • the extent of sea ice would diminish • coastlines would experience more fog and snow
Biological	• a northward shift of the tree line would reduce the arctic ecozone by 15–20%, such that arctic tundra vegetative communities would be restricted mainly to the arctic islands • a northward shift of the tree line would increase competition among the mainland barren-ground caribou herds for preferred calving territory, with potential negative consequences for herd populations • arctic charr and other cold-water fish species would be affected as lake temperatures increased, resulting in the northward expansion of southern fish species, such as brook trout, which compete with the current populations • ocean warming and pack ice recession may increase the range and numbers of some marine mammals, such as beluga and bowhead whales, harbour and harp seals, and walrus • polar bear and ringed and bearded seals require expanses of ice cover for breeding, feeding, and other habitat functions and may suffer population decline through pack ice recession • increased snowfall and warmer winter temperatures (mean below 0°C) would form a snow crust such that the sparse tundra vegetation would be beyond reach of caribou and muskox populations • the Arctic is a primary western hemisphere breeding and moulting ground for shorebirds and waterfowl, and their low-lying coastal habitat could be affected by permafrost degradation and sea-level rise, which would lead to saltwater intrusion • increases in the extent and duration of open water between the arctic islands would limit the movements of caribou, arctic fox, wolves, and other land animals, thereby reducing their opportunity to find suitable habitat and new sources of food.
Socioeconomic	• melting permafrost could damage roads, buildings, and other human-made structures, and onshore oil and gas terrestrial development could become more difficult and expensive • fish and wildlife habitats, upon which the intensity of hunting, trapping, and fishing in an area is now largely determined by accessibility from communities, would be altered • reduction in the extent and duration of sea ice could economically benefit offshore hydrocarbon development, tourism, recreation, and marine transport, as the shipping season is expected to lengthen by six to eight weeks • the Northwest Passage could become a viable shipping route during the summer, although rougher seas, increased by fog occurrence, and more icebergs may occur with changes to the marine environment • a rise in sea level would have serious implications for over half of the arctic communities, which are located on the coast essentially on flat land at or near sea level, extensive and expensive remedial protective measures would be necessary to protect them from flood drainage

Source: Gates *et al.* (1986); Harrington (1986); Sheehy and Chouinard (1989); Hammar (1989); Maxwell and Barrie (1989); Egginton and Andrews (1989); Anderson and Reid (1989, 1991); Intergovernmental Panel on Climate Change (1990).

TABLE 15.7

International agreements and conventions

Agreement	Date
Agreement to Protect Whales	1946
Convention on the Prevention of Marine Pollution by Dumping Wastes and Other Matter	1972
International Agreement on the Conservation of Polar Bears	1973
Convention for the Prevention of Pollution from Ships	1973
Convention for the Prevention of Marine Pollution from Land-based Sources	1974
United Nations Economic Commission for Europe: Convention on Long-range Transboundary Air Pollution	1979
United Nations Convention on the Law of the Sea	1982
Canada/Denmark/Greenland Marine Environmental Cooperation Agreement	1983
United Nations Programme — Montreal Guidelines for Protection of the Marine Environment Against Pollution from Land-based Sources	1985
Convention for the Protection of the Ozone Layer	1986
Canada/Norway Exchange of Letters on Cooperation in Science and Technology	1986
Canada/U.S.A. Porcupine Caribou Agreement	1987
Canada/U.S.S.R. Agreement on Cooperation in the Arctic and the North	revised 1989
The Finnish Circumpolar Initiative on Protection of the Arctic Environment	1991

Discussion for Arctic Pollution

1. Pollution in the Arctic demonstrates that industrial contaminants move freely across national borders, and that some of the Earth's more serious environmental problems – such as ozone depletion and the bioaccumulation of toxic poisons – are truly international responsibilities. Discuss three or four other environmental problems that are having a worldwide impact.

2. The potency of toxic chemicals grows as they move up the food chain. Why is this problem especially acute in the Arctic?

3. Resource industries like oil drilling and mining are inherently "unclean" and damaging to the environment, yet some oil and mining companies are more environmentally responsible than others. How do you think they minimize the damage?

4. The end of the Cold War and the dissolution of the Soviet Union into a number of separate countries has put a halt to the dumping of nuclear wastes in the Arctic Ocean and Baltic Sea. Why? And why did the dumping occur in the first place? Do you think such dumping might resume? What conditions will have to be in place in the new federated states of the former Soviet Union to ensure that radioactivity is more safely contained?

5. Consider the term "sustainable development." Business and governments are fond of the term because it suggests that industrial growth can be maintained but under certain environmental constraints. Others insist "sustainable" and "development" are contradictory; that our environment will eventually become exhausted and polluted if we continue to develop. What do you think? Can large corporations be trusted in the Arctic environment, after abusing it for years? If not, what about all the jobs they represent? How far can companies be regulated before they close down their operations and throw people out of work? What's the solution?

6. Many people believe that the world's most serious environmental problem is the drastic thinning of the protective ozone layer above the Arctic (and the Antarctic). How might steadily increasing levels of ultraviolet radiation affect people, animals, and other species in the Arctic and the rest of the Northern Hemisphere? When environmental damage is caused by contaminants from many thousands of sources across the world, prevention is difficult. But what if only a few industrial companies are primarily responsible? What should be done? Why do you think these companies are still being allowed to produce and sell chemicals that are causing ozone layer destruction?

Introduction

Toxic Pollution in the Great Lakes

This issue examines how the Great Lakes, the world's single most important freshwater resource, has become polluted with toxic chemicals that threaten the survival of many animal species and the health of people who live within the Great Lakes basin.

The Great Lakes are taken for granted by most of the 35 million people who live in proximity to their shores, but the fact is there's nothing like them anywhere on Earth. They are the world's largest freshwater lake system; if you include their tributaries, they account for nearly one-fifth of the world's supply of fresh water; and they are an extraordinary transportation system that ushered world trade into central Canada and the U.S. states of Minnesota, Wisconsin, Illinois, Michigan, and Ohio.

But in taking the Great Lakes for granted, those states and the province of Ontario have managed to pollute the lakes to a degree that is probably irreversible.

The Great Lakes' first environmental tragedy occurred around the turn of this century when epidemics of typhoid and cholera resulted from raw sewage being dumped too close to drinking water intake pipes. (Astoundingly, the city of Sarnia, Ontario pumped its untreated sewage only 45 metres from the conduit through which it drew its drinking water.)

It wasn't until the 1960s that the modern industrial society caught up with the lakes. Decaying algal "blooms" – huge mats of algae nourished by domestic phosphate detergents, human waste, and inorganic agricultural fertilizer – caused a choking depletion of oxygen in Lake Erie. Fish died. Shorelines were fouled by the slimy matter. At that same time, industrial effluents and municipal wastes were dumped into the lakes with such impunity that in one notorious incident the surface of the Cuyahoga River near Cleveland, Ohio actually caught fire.

Today, the beauty of the Great Lakes has largely been restored through restrictions on industrial phosphorous dumping, the reduction of phosphates in detergents, and containment of the most evident examples of corporate and municipal waste disposal.

But the Great Lakes now face a less visible, more deadly scourge: toxic chemical pollution. Chemicals such as PCBs, dioxins, furans, and various pesticides are causing cancers and reproductive abnormalities in wildlife and fish. It is likely they are also harming people who regularly eat the fish. And eventually, they may threaten the health of the 8 million Canadians and 27 million Americans that draw their drinking water from the Great Lakes.

As in other ecosystems, it's the buildup of toxic chemicals that poses long-term risk. In a process known as "bioaccumulation" or "biomagnification," toxins that do not easily degrade collect in plants and animals that exist at a low level in the food chain. These plants and animals are ingested by creatures farther up the chain, magnifying many times in concentration with each predator. In the Great Lakes basin, top predators like cormorants and herring gulls have been found to contain levels of contaminants that are 25 million times as concentrated as they are in the water. Sixteen top predator species in the basin show clear signs of genetic damage due to biomagnification.

Pollutants find their way into the lakes mainly through surface water, groundwater, and the air. Great Lakes tributaries are generally short, which means contaminants reach the lakes rapidly. Furthermore, the large surface area of the lakes makes them vulnerable to atmospheric pollution that can travel thousands of kilometres before falling into the lakes. And they are slow in flushing themselves – the Great Lakes system has an average retention of 191 years. In addition, many Canadian and American cities have aging sewer systems that can overflow during heavy rains, causing raw wastes to bypass treatment plants and flow directly into the lakes.

An enduring and unsolvable problem that will affect the Great Lakes for many years to come is toxic chemicals leaching from hazardous waste dumps and contaminated land. Sites on the U.S. side along the Niagara River – including the infamous Love Canal – are the single largest source of groundwater-borne toxics entering the lakes. There are over 15,000 wells, wetlands, fields, and landfills in the basin where chemicals have been dumped and could be leaking.

Contrary to what many people believe, industrial companies still legally dump toxic waste directly into the lakes and their tributaries. In 1989, for example, Eastman Kodak discharged about 17 million pounds of toxic chemicals from its Rochester manufacturing facility into the Genesee River, which drains into Lake Ontario. Kodak is the fourth largest discharger of toxics into Great Lakes water and the largest discharger of known carcinogens.

Factories like this are known as "point" sources of chemical pollution. There are also thousands of "non-point" sources around the Great Lakes such as toxic rain, oily runoff from city streets, and agricultural runoff. Every year, farmers apply about 33,000 tonnes of pesticides and herbicides to cropland in the Great Lakes basin. A percentage of these toxic chemicals reach the lakes through runoff and seepage. They eventually end up in the food chain.

The governments of Canada and the United States co-operate on Great Lakes issues through an organization called the International Joint

Commission, and jurisdictions bordering the lakes have their own individual and joint cleanup projects. Michigan, for example, has promised to work toward "zero discharge" of toxic persistent chemicals into the system. But environmentalists say that governments are not devising programs that will clean the lakes; they are spending too much time and money monitoring pollution and asking for voluntary toxics-reduction programs. "Only by changing production processes so that we do not use, generate, or discharge these substances will zero discharge be achieved," write Gayle Coyer of the National Wildlife Federation and John Jackson of the Ontario Toxic Waste Coalition.

Ultimately, the pressure to clean up the Great Lakes will have to be exerted on politicians by the people who live within the basin. And until the general public feels the effects of the contamination directly, that's unlikely to happen. In the meantime, the same circumstances that devastated Great Lakes cities 100 years ago unfolds today – waste is dumped in the lakes and drinking water drawn out. The process is just slower and less visible.

Questions and Concerns

READINGS 1 & 2

1. There is a common belief that the Great Lakes have been "cleaned up" since the 1960s when Lake Erie was plagued by rotting algae and dying fish. Why is this not true? Think about the word "clean" and how we define it with regard to environmental pollution.

2. As you read the first reading, list the sources of contamination which find their way into the Great Lakes. Compare this with later readings about government programs to clean the lakes. Is enough being done?

Great Lakes clean-up at critical point

By Peter Gorrie

ON A SUNNY, breezy July morning in Hamilton Harbour at the west end of Lake Ontario, Capt. Doug Greenway was trying to find the spot where three technicians aboard the *Agile*, Environment Canada's research boat, were to conduct their next water tests. He was having trouble because a large freighter anchored in the middle of the bay was blocking the signal from one of the onshore navigation transmitters.

Finally, after circling, he manoeuvred into position and Ken Hill, Bruce Gray and Jacqui Milne got to work, taking samples from the murky harbour and lowering a Sonde water analyser from the stern. At the end of the day, and 25 sampling locations later, their work had produced dozens of pages of charts and graphs and boxes of test bottles to be added to the mountain of data accumulated by scientists at the nearby Canada Centre for Inland Waters.

The good news is that the harbour water is getting better, Hill notes as he watches a graph materialize on an on-board computer from readings relayed up from the Sonde. But the electronic prognosis from below is not entirely encouraging. At least one problem area has been pinpointed in the harbour — a 1.5-square-kilometre section showing virtually no dissolved oxygen. This condition likely means too much ammonia is still entering the water from nearby steel industries and sewage treatment plants.

The day on the *Agile* was like many this crew and others have spent gathering information about the state of the Great Lakes after two centuries of pollution and assessing recent efforts to undo some of the damage.

The findings to date can be summed up as: better, but still being damaged and a long way from recovery. That is the verdict of at least four major reports issued during the past year by private and government organizations.

"Despite the significance of the Great Lakes and our collective rhetoric to restore and enhance them, we as a society continue to mortgage their future by poisoning, suffocating and otherwise threatening them, because of insufficient knowledge, other priorities and short-sightedness," stated one of those documents produced by the traditionally cautious Canada-United States International Joint Commission. In essence, what we are doing to the Great Lakes, we are doing to ourselves and to our children.

The lakes — Ontario, Erie, Huron, Michigan and Superior — are indeed crucial in both environmental and economic terms. They hold one-fifth of the earth's surface freshwater and provide drinking water for eight million Canadians and 27 million Americans. Their basin is home to many animal, bird and fish species. They also support multi-billion-dollar tourism and fishing industries. The transportation routes, hydro power, and water for industrial and human consumption provided by the lakes helped create North America's prosperous industrial and agricultural heartland.

But pollution, development, and destruction of wildlife habitats have pushed some Great Lakes species to extinction and continue to endanger others. In many places, the water is still a threat to human health, mainly because it contains toxic chemicals that accumulate in the flesh of fish and can cause reproductive problems for those who eat them. Swimming is prohibited frequently during the summer in parts of the lakes touching on such large urban centres as Toronto.

In 1615, Samuel de Champlain described Lake Huron as a "sweet sea," but it and the other Great Lakes did not remain sweet for long. After European settlers began flocking to the shores about 200 years ago, the lakes became a convenient dump for human effluent as well as wastes from farms, mines, mills and the region's growing manufacturing industries. For decades, the consequences were ignored, even when it became clear that the lakes, though immense, were not too big to be affected by humans. Then, a series of crises prompted some limited action to clean up the worst of the mess.

At the turn of this century, cholera and typhoid epidemics swept the burgeoning cites of the Great Lakes, killing hundreds of people. It was the inevitable result of dumping raw sewage into the lakes and then using the untreated lake water for drinking. Most cities responded by chlorinating the drinking water to kill the bacteria in it, not by reducing pollution. The deaths stopped and so did public interest.

In the mid-1950s came warnings that Erie — the shallowest and warmest of the lakes — was threatened by phosphates and other nutrients contained in detergents, particularly from sewage treatment plants and fertilizers in farm run-off. At first the alarm was met with skepticism, but

Courtesy of Peter Gorrie

before long there could be no debate about the problem. The lake was so over-enriched it was said to be choking to death on massive blooms of bright-green algae that covered parts of its surface in mats up to a half-metre thick. As the blooms died and decayed, they consumed the waters' life-giving oxygen.

Large pieces of the mats also drifted ashore, fouling beaches and blocking water intakes in a mess made even more putrid by the addition of detergent suds and rotting fish. By the late 1950s, the public outcry had forced governments — over the protests of some business groups — to restrict the use of phosphates in detergent and led to construction of about $10 billion worth of sewage treatment plants around Erie's shores, as well as measures to control farm run-off.

But as Erie's waters were being returned from green to blue, another problem, more dangerous and difficult to contend with, was uncovered. All five lakes — even cold, remote Superior — were found to contain toxic chemicals. In the late 1960s, the first links were made between some of those chemicals and human health problems. At that time, the focus was on cancer.

The discoveries, coming at a time of widespread concern about the impact of DDT and other pesticides, were so alarming that politicians again felt compelled to act. Canada and the United States banned or restricted 11 chemicals believed to cause cancer or birth defects, including DDT, the pesticides mirex and toxaphene, and polychlorinated biphenyls (PCBs). In 1977, the United States enacted the Clean Water Act requiring polluters to start reducing toxic discharges. Ontario, the only Great Lakes province, half-heartedly followed the American lead by setting pollution guidelines. In 1978, the United States and Canada signed an agreement that set as its goal "zero discharge" of toxic wastes.

The regulations were not stringent, and enforcement by government agencies was often lax, but the measures succeeded, for a while, in cutting concentrations of toxic chemicals in the water. Since the early 1980s, however, pollution levels have reversed and a growing number of substances, including many that are highly toxic and banned in Canada, have been detected in the lakes. The worst of these chemical contaminants, such as dioxins and furans, cannot be seen, smelled or tasted. In fact, their presence can only be detected by highly calibrated equipment capable of measuring such minute quantities as one part per quadrillion.

It is hard to pin down exactly what this pollution is doing to humans. During the cholera and typhoid epidemics people died quickly and there was no doubt as to the cause. In later years, the stinking piles of algae and rotting fish piled up along lake Erie beaches clearly nauseated people. Here was visable evidence of pollution at its ugliest. Today, cosmetics have transformed the lakes and restored much of their beauty. But it is a deceiving metamorphosis indeed. Beneath their seemingly pristine surfaces lie an insidious mixture of toxic chemicals.

To date, most of what we know of the effect of the chemicals on human health has been inferred from studies of bald eagles, herring gulls, snapping turtles and other wildlife. Only a few projects have looked directly at human health effects. Their results, though considered significant, have not been frightening enough to produce the level of public outrage needed to force business and government to act decisively.

As a result, clean-up and research budgets are underfunded, Canadian and American governments have failed to achieve the zero discharge goals, new pollution-control legislation appears riddled with loopholes, and some Great Lakes states — particularly Wisconsin — are even relaxing their environmental regulations to attract new industries.

"Great Lakes research and management are at a critical juncture," according to the Washington-based Northeast-Midwest Institute, a non-profit research group. It complains that federal funding for Great Lakes programs in the United States has dropped 11 percent since 1981. "Funding reductions not only threaten future programs but may undermine progress already achieved."

Toxic pollution is a consequence of the chemical revolution following World War II that created plastics, solvents, pesticides, cleaners, fuels and a myriad of industrial products. About 65,000 such substances are manufactured in North America, and traces of nearly 1,000 have been found in the lakes, along with heavy metals such as mercury, lead and cadmium.

They have come from many sources, both point and non-point. Hundreds of pulp mills, steel factories, petro-chemical complexes and auto assembly plants are point sources pouring waste water directly into the lakes or their tributaries, with varying amounts of pollution control. Pollutants include organic compounds such as dioxins, polycyclic aromatic hydrocarbons and methylene bromide; poisonous metals such as mercury, lead and chromium; and nutrients such as ammonia that contribute to oxygen depletion in water.

Other point sources — an estimated 11,000 in Ontario's Great Lakes basin and tens of thousands more on the American side — are smaller industries that dump toxic wastes into municipal sewer systems, where they join the mixture of solvents, paints, oils and cleaners that individuals pour down household drains. Most of the chemicals run straight through sewage treatment plants, which are not designed or equipped to remove them from the waste stream. Some even kill the bacteria used to break down conventional sewage wastes.

The problem is compounded by outmoded sewage systems under many Canadian and American cities in the Great Lakes basin. In the complex networks of pipes, it is almost impossible to pinpoint pollution sources. To make matters

Exotics push out native species

THE INVASION began in 1988, when a ship that had crossed the Atlantic carrying water as ballast emptied its holds into Lake St. Clair, in preparation for taking on cargo.

The water it dumped contained zebra mussels, tiny creatures that were common in European lakes and rivers but had never been seen in North America. These bivalves with yellow and brown stripes made themselves right at home in their new environment. In fact, with abundant food and no major predators, their numbers exploded and they began spreading like an underwater plague.

Before long, these innocuous-looking invaders were known and bemoaned throughout the Great Lakes basin. By the summer of 1989, the Detroit River had carried them into Lake Erie where they flourished. Now they have been swept farther downstream through the Niagara River into Lake Ontario and on into the St. Lawrence River. Others have hitched rides on boats travelling upstream to lakes Huron, Michigan and Superior.

Wherever they go, the mussels, which can grow to a length of only four centimetres but are usually much smaller, are doing enormous damage to both the natural environment and man-made installations. Getting rid of them is considered an impossibility, and Ontario's Ministry of Natural Resources estimates it will cost more than $1 billion just to make co-existence possible.

Zebra mussels are only the latest of more than 100 species that have been introduced — some accidentally — into the Great Lakes as a result of human activity during the past 200 years. These so-called exotics are among the reasons the lakes' aquatic life has been transformed since Europeans began settling on these shores.

In their natural state, the lakes teemed with sturgeon up to two metres long, Atlantic salmon, lake trout and whitefish that all thrived at the top of a rich, complex food chain.

But the new species, combined with pollution, overfishing and destruction of wetlands, gravel beds and other vital habitats, have destroyed the old order. The lakes are now dominated by coarse fish such as carp, alewives, perch and smelt, as well as the predatory lamprey eel, which decimated the remaining trout and whitefish after swimming up ship canals from the Atlantic. Ironically, the lakes are now being stocked with two varieties of Pacific salmon — which cannot reproduce in these waters but do well there after being supplied from hatcheries — to create a sports fishing industry.

Zebra mussels are not the only recent invaders: in 1987, the northern European river ruffe was found in Duluth Harbour on Lake Superior, causing concern because it has no commercial value and can dominate perch and whitefish by taking over their spawning beds and those of other species.

About 1,000 ocean-going ships dump millions of litres of ballast water into the lakes every year. Unless the practice is stopped, more exotics are certain to arrive.

"This kind of biological pollution is a permanent change; unlike a chemical or oil spill which will go away in 100 years or so," warns Margaret Dochoda of the Great Lakes Fishery Commission in Ann Arbor, Mich. "Also, it isn't localized, it spreads."

Zebra mussels have the potential to be among the most damaging of the new arrivals. Each female produces up to 40,000 eggs annually. Microscopic larvae, or veligers, emerge and grow for a summer, then attach themselves to a solid surface to spend the rest of their five-year lifespan. If the mussels have covered all available surfaces, new arrivals settle atop those already attached, quickly building up a thick crust.

The bivalves coat commercial fishing nets, not only making them ineffective but also wrecking the machinery that winds them back onto fishing vessels. They can, within a matter of weeks, cover smooth boat hulls, and their weight has sunk a number of navigation buoys. They stop the flow of water through vessels' cooling systems, causing overheating and creating the risk of fire. They clog the intake pipes of water treatment facilities, nuclear power plants, industries and farm irrigation systems.

The mussels also settle on rocky lake bottoms where pickerel and other fish spawn. Eggs have been deposited on top of the mussel colonies, but it is not clear yet whether this new environment is hospitable.

Adult zebra mussels feed by filtering plant material out of the water. They are extremely efficient at this, and they have already removed so much matter from the once-murky western basin of Lake Erie that its water clarity has increased by 80 percent, March says.

That sounds like a positive impact, but they are consuming a valuable food source for others in the food chain. As well, some fish, including pickerel, cannot tolerate the sunlight now shining through the clarified water.

Some controls are possible. The veligers — often found in concentrations of 30,000 per cubic metre of open water — are too small to be stopped by screens on intake pipes. But zebra mussels dislike deep water and so may be avoided if intake pipes are extended farther into the lakes. However, most of Lake Erie and all of Lake St. Clair are too shallow for that strategy to work. And lengthening pipes is an expensive proposition in the deeper lakes. In Europe, one of the main control measures is to build double water intake pipes: one is used when mussels are being cleaned out of the other. But that, too, would be extremely costly.

The mussels can be killed by flushing hot water through the system, but that would require major design changes, and could cause thermal pollution if that water is pumped into the lake before being cooled. Chlorine is also an effective flushing agent, but it also poses a pollution threat; when combined with organic matter, it can create cancer-causing trihalomethanes

"There is no universally accepted method for zebra mussel control," concludes a report by the Ontario Ministry of the Environment.

However, research is underway in Toledo, Ohio, into the use of an extract from a common African plant, the endod, or soapberry, as a weapon against the mussels. The plant's active ingredient, oleanolic glucocide — now called Lenima toxin – is a natural toxin, water soluble and would not build up in the lakes. However, while deadly to zebra mussels, it is also toxic to fish and other organisms at some levels of concentration.

In four or five years, it is expected the mussels will run out of food. They will then suffer a major die-off, which will foul shorelines with billions of their decaying bodies. After that, their numbers may stabilize at a lower level, but they will persist, adding a permanent new dimension to life in the Great Lakes.

To prevent new exotics from arriving, Canada and the United States would have to pass laws that require ocean vessels to dump their ballast water well out in the gulf of St. Lawrence and replace it with salt water if necessary, Dochoda says. Any life forms picked up in the gulf would be unlikely to survive in fresh water, and the Great Lakes would not be harmed by the salt water that would be dumped into them.

Both governments have expressed interest and issued advisories to the shipping industry but, to date, ballast exchange is only voluntary.

Peter Gorrie

worse, sanitary and storm sewers are often connected. After a heavy rain, run-off waters can overtax treatment plants, allowing human and chemical wastes to by-pass treatment facilities.

Many of the point sources have been drastically reduced. At Sarnia, for example, a Dow Chemical Canada Ltd. complex, which five years ago was blamed for creating a massive toxic blob on the floor of the St. Clair River, has eliminated 99 percent of its waste discharges through new industrial processes and containment. While some cities are separating storm and sanitary sewers, or building huge holding tanks to store excess storm water until it can be treated, others say they cannot afford to take those steps.

In 1986, the Ontario government launched the Municipal-Industrial Strategy for Abatement (MISA) intended to identify and limit every source of industrial water pollution in the province. In theory, it is an improvement over the old guidelines that often allowed industries to simply dilute their pollutants, instead of limiting the actual amount they could discharge.

But watchdog groups such as Pollution Probe and the Institute for Research on Public Policy argue that after four years, MISA has not stopped a single drop of polluted water from entering the Great Lakes. The strategy only requires industries to install the best available pollution-control equipment they can afford, the groups claim. It has not set absolute water-quality standards that would require development of better clean-up technologies.

In the United States, most industries and municipalities have cut pollution below the maximums allowed by state and federal laws. But those regulations, too, are not stringent enough to produce acceptable water quality, says George Coling, a Great Lakes specialist with the Sierra Club.

The clean-up has been inadvertently aided by the decline of some manufacturing industries. Giant steel mills along Buffalo's Lake Erie shoreline are being replaced by parks, marinas and condominiums.

Less progress has been made on reducing non-point sources of pollution. More diffuse and often hidden, they are much harder to contain.

One of them is air pollution from within the Great Lakes basin and from around the world, which creates toxic rain. The Sierra Club recently described it as the largest unregulated source of new pollution in the lakes.

An estimated 90 percent of the PCBs in Lake Superior and 30 percent of those in Lake Ontario are believed to come from air pollution, probably the result of evaporation when chemicals in that family of chlorinated hydrocarbons are in use, spilled or improperly stored. While PCBs are no longer being produced in North America, their past use in a range of products from waxes to paints and pesticides has left an environmental legacy that may linger for decades. Fresh deposits of DDT also are being detected apparently because the pesticide is being carried through the air from Central and South America, where it remains in heavy use. Fish from small lakes in Wisconsin, Minnesota and northern Michigan — all in the Great Lakes watershed — cannot be eaten because of contamination by airborne mercury.

American politicians, however, have now passed an updated Clean Air Act to impose tougher emission controls on cars and smokestacks. Among other things, the act is aimed at reducing emissions of almost 200 toxic chemicals by 90 percent over the next 10 years.

From the land, pesticides and animal wastes are washed into the lakes in the millions of tonnes of soil eroded each year. Programs have been launched in Ontario and the United States to reduce the use of farm chemicals and control erosion, but most are voluntary and have had only limited impact.

Even if these sources were shut off completely, the pollution would not be stopped. Past mistakes would still haunt the lakes.

The basin is dotted with at least 15,000 wells, wetlands, fields and landfills where chemicals have been dumped. Many of these pools of toxic chemicals are leaking into the ground water and seeping toward the lakes. The most famous of these are Love Canal and other waste dumps around Niagara Falls, N.Y., whose contents were found to be oozing through the rock face of the Niagara Gorge and contaminating the river and Lake Ontario — the most chemically polluted of the five lakes. The seeping waste included the most toxic form of dioxin, 2, 3, 7, 8 TCDD, among the deadliest of all man-made substances and used mostly in wood preservatives and pesticides.

After a long court battle, U.S. government agencies backed down from insisting the Niagara Falls dumps be excavated and agreed instead to try to contain the leaking chemicals.

Governments in both countries have passed laws to control the handling and dumping of hazardous wastes, but that has not stopped illegal dumping.

Sixteen abandoned waste dumps on the U.S. side of the Great Lakes are "superfund" sites — money for cleaning them is to come from a $9-billion fund created by taxing industries generating toxic wastes. But the current fund is not big enough to cover all the sites and, as the Sierra Club and Northeast-Midwest Institute complain, some will not be dealt with for years.

Yet another threat to the lakes are the toxic chemicals that have settled into the lake-bottom sediments in many areas, particularly in harbours and at industrial outfalls. When waves, currents, dredging, large ships or even feeding carp disturb the sediments, the chemicals they contain are recirculated into the water.

The International Joint Commission — established in 1909 by Canada and the United States to deal with Great Lakes water use and diversion issues — has designated 42 areas of concern in the Great Lakes requiring urgent attention. All but one include grossly contaminated sediments. Worst is the harbour at Waukegan, Ill., on Lake Michigan, where fully half of the sediments are PCBs from an Outboard Marine Corp. small-engine plant. After a decade-long court battle, the company has agreed to pay for a $21-million clean-up. Hamilton Harbour remains the most contaminated area on the Canadian side, despite a concerted pollution-control effort by the city's steel mills (*CG* June/July '87).

Contaminated sediments can be removed, and the U.S. government is funding some trial runs. But the cost of dredging all the seriously polluted sites in the lakes could be as high as $40 billion, and the effort would likely create new problems. The chemicals would be mixed with the water during dredging. Just as serious, few sites have been found yet to safely dispose of the millions of tonnes of contaminated material that would be scooped out of the lakes. Besides, scientists say, there is little point in dredging if continuing pollution will contaminate the sediments again.

At each of the 42 areas of concern designated by the International Joint Commission, local groups have been established to draw up remedial action plans. After five years of work, they have produced a lot of meetings but few results. They have no deadlines, each group sets its own standards for desired water quality, and there is no guarantee of clean-up funding.

The cost of removing sediments, curbing erosion, cleaning up waste dumps, reducing pollution and taking all the other steps needed to create lakes that are not a threat to humans and wildlife is likely to total at least $100 billion.

Estimates are unreliable because, as the Northeast-Midwest Institute notes, "scientists have only begun to understand the full extent of the damages caused by toxic contamination...(and) engineers are still in the early stages of reviewing the developing (clean-up) technologies."

When all the sources are taken together, it presents a very complex problem. "We don't exactly know what has to be done to clean up the place properly and get out of the mess we've created over 40 years," says Mike Gilbertson, a researcher with the International Joint Commission.

But environmental groups and many scientists contend that even though opinion polls suggest the environment is a major concern, governments, industries and consumers will not be willing to pay the price until there is clear evidence of a human health risk.

"People will have to feel it before they do it," says Jack Vallentyne, a senior scientist at the Canada Centre for Inland Waters in Burlington, Ont., and Canadian co-chairman of the International Joint Commission's Science Advisory Board. He is convinced that the danger "is beginning to hit home." But groups such as Greenpeace and the Sierra Club, as well as other scientists, are not so sure, because the contamination has not had an obvious impact on most people in the Great Lakes basin. They do not appear to get sick from breathing the air or drinking treated water from the lakes, and the danger appears remote and theoretical.

At most risk, because of a process called biomagnification, are those who eat fish taken from the lakes.

The chemicals of greatest concern break down slowly in the environment and accumulate in the tissues of living things. The concentration builds dramatically at each level of the food chain. In top predators such as cormorants and herring gulls, the level of contamination may be 25 million times as high as it was in the water.

Humans who eat trout, pickerel or other fish near the top of the lakes' food chain thus get a concentrated dose of chemicals. The exposure is considered dangerous enough that governments warn against eating certain species.

Studies have produced clear evidence that toxic chemicals have damaged 16 top predator species in the Great Lakes basin. Some, such as eagles, are unable to reproduce in certain areas; other species are badly diminished. Gulls lay eggs with shells so thin they break under the parents' weight. Some chicks are born weak and undersized because they could not absorb the nutrients in their egg yolks, or suffer from edema, a condition in which liquids accumulate under the skin and around the heart. Some cormorants are born with decaying livers, or with beaks so deformed they cannot eat.

Many scientists assume the chemicals could have similar effects on humans who eat a lot of Great Lakes fish. For years, public concern centred on cancer, because it was the focus of scientific studies, and media reports invariably described the chemicals in the lakes as "suspected carcinogens" or, at least, "linked to cancer in laboratory animals." However, researchers are shifting their emphasis in response to evidence that toxic pollution may alter human reproduction.

Surprisingly, only two studies have tried to find evidence of defects in humans. Their results are inconclusive, but ominous:

At Wayne State University in Michigan in 1981, researchers began comparing 242 pregnant women who ate fish caught in Lake Michigan with 71 who did not. The fish eaters — who consumed, on average, just 450 grams of lake fish per month — gave birth to babies with lighter bodies and smaller heads. When given neuromuscular tests at four years of age, these children also exhibited slower nervous-system development. The changes increased in proportion to the amount of PCBs in the mother's blood.

The second study, at the University of Wisconsin, concluded that children of women who ate fish from Lake Michigan contracted more infectious illness during their first four months than did children of a control group.

More research is needed to confirm these results and "it is a disgrace that no agency has attempted to do so since the initial findings were published in 1984," scolded U.S. scientist Ian Nisbet at a joint commission workshop in Chicago last year. Studies have not even been done on those considered most at risk because they eat a lot of fish — native people on the Walpole Island Reserve on the St. Clair River, and others at the east end of Lake Ontario and along the upper St. Lawrence River.

Nisbet's advice, echoed by many other scientists and environmentalists, is finally being acted on by Health and Welfare Canada, which has embarked on a long-term study of the health of Canadians in the Great Lakes basin.

If the results are similar to those of U.S. studies, pressure will mount for a massive clean-up.

Environmentalists and scientists warn that if the aim is to get close to zero discharge of industrial waste, it will not be enough to simply spend billions of dollars to plug the pollution sources, because there are limits to what that can accomplish. The main solution, they say, is prevention; taking steps to cut the amount of chemicals produced and used in the Great Lakes basin, and elsewhere.

Consumers will be required to alter their lifestyles, industries will need new production processes and farmers will have to break their dependence on chemical pesticides and fertilizers. "Some chemicals we just have to do without or we'll never get off the toxic treadmill," says Paul Muldoon, a lawyer with the Canadian Environmental Law Association.

"The Great Lakes are a symptom of what we've allowed our society to become," says Andy Gilman, a toxicologist who is heading the federal government's health study. Pouring money into a clean-up effort will not solve the problem, he says. "It requires a change in individual and corporate attitudes. We're all part of the problem. People have to stop pointing fingers and decide what they're going to do about it." ❖

Peter Gorrie is a Toronto journalist specializing in environmental subjects.

Toilet bowl current stops swimmer

VICKI KEITH says she has never seen Lake Ontario as polluted as it was during a triple crossing attempt last August.

A crew member pulled her from the water 11 kilometres from the end of the first leg of the journey. Keith's ears hurt and she had vomited repeatedly during the previous three hours.

She emerged from the lake with a layer of scum on her swimming cap and goggles. The water had turned her white bathing cap yellow and an ugly crust had formed on her bathing suit. Dead seagulls were floating on the lake and one was pulled directly from her path.

Keith, 29, had chosen a route farther east than the one she normally swims, taking her through a "toilet bowl current" of raw sewage. She began at Niagara-on-the-Lake and was headed toward the Leslie Street Spit at Toronto.

The Toronto native, who holds 17 world swimming records, has been swimming in the lake for 11 years and says this was the worst she has seen it. When Keith speaks to students and other groups her concern about pollution sometimes rivals her passion for swimming. She says people need to know about the state of the lake and the need to start cleaning it up. "I have never seen anything that disgusting," she says.

Courtesy of Peter Gorrie

Reading 3

Questions and Concerns

READINGS 3 & 4

1. Examine the language and tone-of-voice used by the government officials in their description of cleanup programs for the Great Lakes. Compare it to the language and tone-of-voice used by the environmentalists in their rebuttal. What are the differences? Reading between the lines, what does this tell you? Consider the importance of language when discussing environmental issues.

Lake Superior and Zero Discharge
The Effort By Government

by Chuck Ledin, Wisconsin DNR, and Jake VanderWal, Ontario Ministry of the Environment

In its Fifth Biennial Meeting in Hamilton, Ontario in 1989, the International Joint Commission (IJC) recommended that, "The parties designate Lake Superior as a demonstration area where no point source discharge of any toxic chemical will be permitted."

In response to this challenge, the governments of Canada and the United States formed a Task Force of Senior Managers to outline "A Binational Program to Restore and Protect the Lake Superior Basin." This program reflects a strong commitment of the governments of Canada and the U.S., the states of Michigan, Wisconsin, Minnesota, and the province of Ontario to restore degraded areas of the basin and protect this Great Lake for future generations.

In the early development of this program it was recognized that dealing with toxic chemicals alone, as specified under Annex 2 of the Great Lakes Water Quality Agreement, would limit the effectiveness of any efforts to protect the Lake Superior ecosystem in the long term. Thus, a broader, more extensive and comprehensive approach to managing the watershed evolved. The Lake Superior Binational Program was released at the IJC's 1991 Sixth Biennial Meeting in Traverse City, Michigan.

The ultimate goal of this program is to restore and protect the integrity of the Lake Superior ecosystem through the development of creative pollution prevention strategies, enhanced regulatory measures, and, where necessary, remedial programs. The unique character and pristine nature of Lake Superior will be highlighted through a special designation status, and, as immediate progress is made towards virtual elimination of designated persistent bioaccumulative toxic substances, a broader, long-term program will be developed to protect the basin from the release of other toxic substances and from further habitat destruction.

The broader program embodies a holistic approach to understanding and managing this headwater lake. As part of this program, the state of the basin is being assessed and a "vision" is being developed from which goals and objectives will be stated. This information will provide the foundation necessary to complete a long-term, basin-wide management plan. This broader program will include measures to control point and non-point source impacts, habitat management (with special emphasis on rare or endangered species), and activities to reclaim and restore degraded areas. It will also fulfill commitments under Annex 2 of the Great Lakes Water Quality Agreement (revised, 1978). A monitoring and surveillance strategy will be established to track the success of the program.

The Lake Superior Binational Program builds upon the development of partnerships between all

Courtesy of Great Lakes United

levels of government, and incorporates the expertise of industry, municipalities, universities, aboriginal peoples, environmental organizations and the public. To facilitate citizen input, the Lake Superior Forum was created, and has been supported by the parties. This public advisory group has both Canadian and U.S. co-chairs, and a membership of twenty-two individuals representing the various interest groups identified. Secretariat was originally provided by the Lake Superior Centre in Duluth, Minnesota, but is now housed on both sides of the border at the Sigurd Olson Environmental Institute in Ashland, Wi. and Lakehead University in Thunder Bay, Ontario.

The Lake Superior Forum has conducted 31 stakeholder interviews in the basin to identify barriers to pollution reduction and to propose means of overcoming them. An executive summary is available. The Forum has also endorsed a vision statement (January 31, 1991), and has promoted awareness of the Lake Superior Binational Program through public outreach (e.g., Inland Sea Symposium, Bayfield, Wi.; National Council on Air and Stream Improvement, Minneapolis, Mn.; MultiStakeholder Meeting, Sarnia, Ont.). These activities have been supported by the Governments.

The Superior Work Group, comprised of technical and scientific professionals from each of the sixth jurisdictions, is in the process of developing and implementing the Lake Superior Binational Program. A workplan addressing each of the thirty-seven action items in the Program has been drafted. The development of this document has already stimulated binational dialogue and resulted in greater program coordination between Federal, State and Provincial agencies. New initiatives have been proposed, and some existing programs have been expanded or accelerated to meet the IJC's challenge.

Accomplishments to date include:

Pollution Prevention

• An Environmental Community Action Plan program has been launched by Environment Canada and the Conservation Council of Ontario on the Canadian side of the Lake Superior Basin to introduce this program in four focus communities. Community Action Plans provide a framework for increasing the effectiveness of environmental problem solving, by identifying and prioritizing issues, developing a community network and seeking technical and financial assistance.

• Domtar, Inc., Environment Canada and Environment Ontario have initiated a feasibility study of chlorine free bleaching technologies that will be demonstrated at Domtar's Red Rock Mill. This laboratory program will focus on ozone delignification or oxygen/peroxide delignification as a first stage, enhanced by pretreatment such as with enzymes or sequestering agents. Pulp brightening agents such as hydrogen or sodium hydrosulfite will be used to achieve the target brightness for semi-bleaches pulp. Bleaching technologies that approach 80+ pulp brightness levels will be investigated, thereby increasing the number of mills capable of implementing the research findings.

US EPA is conducting an inventory of hazardous waste sites in the Lake Superior Basin and identifying pollutants disposed of, generated, or stored at these facilities. The three basin states are also conducting facility audits to evaluate further reduction opportunities.

• A U.S.-based agricultural Clean Sweep program has been implemented to properly dispose of banned or cancelled pesticides to prevent contamination of the watershed in four Minnesota Lake Superior Counties.

• US EPA has launched a Lake Superior "Mini Exchange" to permit individuals and organizations to access a clearing-house of technical, policy, programmatic, legislative and financial information on efforts to reduce pollution in the area. This information can be accessed by any computer with a modem at (703) 605-1025. For more information, please contact Chris Flaherty, US EPA — Region 5 at (312) 886-4077.

• The Lake Superior Partnership is a joint project between the State of Minnesota and the Western Lake Superior Sanitary District in Duluth to conduct multi-media compliance inspections and identify pollution prevention opportunities. Ten inspections have been undertaken over the past year, each reflecting the type of operation and permits held by that facility. This multimedia inspection, involving air quality, water quality, solid waste, hazardous waste and wastewater pretreatment, demonstrates, at a smaller scale, the holistic approach which is being applied to the Lake Superior watershed. The Lake Superior Partnership is an example of government, industry, small business, academia, and environmental organizations working together to reduce the discharge and emissions of toxic substances into the environment.

• The Western Lake Superior Sanitary District (WLSSD) has conducted fourteen household hazardous waste collections in northeast Minnesota in the last year, and has established a permanent hazardous waste collection site as part of their regular operations. Over the last three years, this facility has been utilized on an average of over 3000 households per year, which is effectively diverting tons of hazardous waste from improper disposal at landfill sites. In addition, the WLSSD has developed programs for collecting nickel-cadmium (rechargable) batteries and button batteries, and is in the process of developing a program for collecting materials containing mercury from small business.

Control and Regulation

- A demonstration project to develop comprehensive stormwater management plans for nine Lake Superior cities in the United States is underway. Monitoring stormwater discharge for the presence of persistent bioaccumulative substances will help to identify sources and lead to greater control. The need for Best Management Practices to control pollution from bulk storage sites along the waterfront in these nine cities will also be evaluated.
- Canada has finalized the Pulp and Paper Regulatory Package to ensure stricter and consistent regulations for pulp and paper mills in Canada, by reducing conventional pollutants and eliminating dioxins. This new legislation sets more stringent effluent limits on BOD and total suspended solids, no acutely lethal effluent, no measurable concentrations of 2,3,4,8-TCDD and 2,3,7,8-TCDF in pulp mill effluent, and a ban on the use of pentachlorophenol treated woodchips by pulp mills. Environment Canada is currently implementing these regulations.
- Ontario's Municipal-Industrial Strategy for Abatement (MISA) is finalizing regulations for the pulp and paper. Negotiations with the pulp and paper sector are currently underway to finalize these regulations. Federal-provincial discussions are being undertaken concurrently to harmonize MISA regulations with recent Canadian Environmental Protection Act and Fisheries Act amendments.
- Lakefilling and Sediment Quality Guidelines were released by the province in June 1992 that will ensure that materials added to Ontario's waters will set safe levels for metals, nutrients and organic compounds.
- In May of 1992 the province released a Candidate Substances List for Bans or Phase-outs, composed of 21 of the most hazardous substances.
- The U.S. initiated a comprehensive lake, tributary and atmospheric monitoring program to further identify sources of persistent bioaccumulating substances. A new International Atmospheric Deposition Network Station has been established on the Keweenaw Peninsula.

Special Designation

- The U.S. will explore the possibility of a Lake Superior Water Trail.
- The U.S. and Canada are exploring the potential of a "binational" Lake Superior Biosphere Reserve as part of the UN's Man and the Biosphere Program.
- The U.S. has proposed a designation system for Outstanding National and International Resource Waters in the Lake Superior Basin under the Great Lakes Water Quality Initiative and Environment Canada is considering a similar designation.

Restoring and Protecting the Lake Superior Basin

- The Superior Work Group will issue four volumes that will identify the agency strategies, chemical threats and impairments, the management challenges and binational, basin-wide management plan that will restore degraded portions of the Basin, and protect the integrity of this Great Lakes ecosystem.
- Remedial Action Plan programs will be accelerated where possible. Contaminated sediment remediation programs are underway at two Areas of Concern (Torch Lake and Thunder Bay), while degraded aquatic ecosystems are being restored at Thunder and Nipigon Bays.
- The current status and loadings of persistent toxic chemicals in air, water, sediment and biota are being collected as part of the State of the Basin Report. Water Quality standards, sediment guidelines and fish consumption advisories will be compared to ambient levels of contaminants. Exceedance of acceptable levels will define "critical" pollutants, and identify where immediate action is necessary to reduce loadings. January 1991 has been established as a baseline from which to measure success.
- The Superior Work Group has identified the need to develop a coordinated and complimentary monitoring and surveillance program to track progress of restoring and protecting Lake Superior. Government agencies are developing standards for identification of chemicals, loadings, use impairments and ecological indicators of environmental health. Over the next five years, existing programs will be integrated and new programs developed to ensure streamlined and efficient reporting of the status of the Lake Superior basin.

As the Binational program develops and unfolds, Lake Superior has its own chapter in IJC's Sixth Biennial Report on Great Lakes Water Quality. The Commission recognized that "over the past two years, no other recommendation has generated more enthusiasm and hard work on the part of governments, non-governmental organizations and individuals to develop such a program."

The program is in its infancy, and as such, it remains exciting and the challenges are great. Success of this program depends on building strong partnerships between government agencies and all other stakeholders. The Parties and jurisdiction have demonstrated a strong commitment to Lake Superior by focusing significant financial and technical resources to build these partnerships and deliver a program to restore and protect this Great Lake. Lake Superior will be the model for the remaining four Great Lakes, and set an example of global proportions!

The authors co-chair the Superior Work Group for a Binational Program to Restore and Protect the Lake Superior Basin.

Lake Superior and Zero Discharge
Environmentalists Respond

by Gayle Coyer, National Wildlife Federation and John Jackson, Ontario Toxic Waste Coalition

The words "zero discharge" are not even mentioned in the governments' article (see Reading 3) on the Lake Superior Agreement. This omission, more than anything else, shows the failure of the governments to live up to the commitment made a year ago to achieve zero discharge in Lake Superior.

The Lake Superior Binational Program was released with much fanfare in Traverse City, Michigan, at the International Joint Commission's Sixth Biennial meeting in October 1991. In the Binational Program, the governments with jurisdiction in the Lake Superior basin —Canada, Ontario, the United States, Wisconsin, Michigan and Minnesota—committed "to achieve zero discharge and zero emission of certain designated persistent, bioaccumulative toxic substances, which may degrade the ecosystem of the Lake Superior basin." Regrettably, the governments are not developing programs that will achieve zero discharge. Instead, they are focusing on monitoring and cautious voluntary toxics reduction programs.

The whole point of zero discharge and the thrust of Lake Superior as a demonstration area for zero discharge is a recognition that we must fundamentally shift how we deal with control measures that have not and will not work. Only by changing production processes so that we do not use, generate or discharge these substances will zero discharge be achieved.

Achieving zero discharge in Lake Superior means examining existing industries on a sector by sector basis, analyzing the wastestreams of the industries in that sector, and determining alternative production processes to eliminate the use and release of persistent toxic pollutants. It means giving special designation in the U.S. and a comparable designation in Canada — to prohibit new or increased sources of persistent toxic pollutants. The designations must be given to the entire Lake and not just little pieces of it, as the governments propose, since pollution knows no boundaries. Bold, innovative programs are also needed to eliminate pollutants that come from the atmosphere.

In the Sixth Biennial Report, the IJC evaluated the Lake Superior Binational Program. The Commission astutely concluded that the Binational Program appeared to be more of a program to reduce and manage persistent toxic pollutants, rather than to eliminate them. The IJC defined for the the governments what zero discharge for Lake Superior should mean: prohibit new or increased sources of point source discharges and phase-out existing sources. The Commission also recommended that the governments establish a specific date by which no point source releases of any persistent toxic substance will be allowed into Lake Superior or its tributaries.

The six governments deserve credit for the unprecedented way they are working together to develop programs for Lake Superior. However, the governments still have not gotten it right — they are not implementing programs that will achieve zero discharge. It is up to the citizens of the Great Lakes basin to continue to pressure the governments until they truly understand zero discharge means changing production processes and until programs to do this are implemented for Lake Superior.

The United States Congress has just appropriated an additional $950,000 for the Lake Superior Binational Program. We call on elected officials to ensure that this money and additional monies from Canada are used for zero discharge programs and not for monitoring and voluntary toxic use reduction programs.

Courtesy of Great Lakes United

Questions and Concerns

READING 5

1. The term "sunsetting" is used in this reading to describe "a process to restrict, phase out, and eventually ban the manufacture, generation, use, transport, discharge, and disposal of a [persistently toxic] substance." Industrial companies that manufacture and use dangerous substances resist this idea, or call for gradual phase outs that might take many years. What are the ethics of their stance? Should sunsetting be voluntary or regulated? Name some chemicals that should probably qualify for sunsetting. Why?

Sixth Biennial Report Emphasizes Virtual Elimination of Persistent Toxic Substances

"Are humans and our environment in danger from persistent toxic substances now? Are future generations in danger? Based on a review of scientific studies and other recent information, we believe the answer to both questions is yes."

by Sally Cole-Misch

Thus concludes the International Joint Commission in its *Sixth Biennial Report on Great Lakes Water Quality*, released in mid-April on the 20th anniversary of the signing of the Great Lakes Water Quality Agreement. The Commission's report focuses on persistent toxic substances, and the steps needed to reach the Great Lakes Water Quality Agreement's goal of virtual elimination of the inputs of these substances to the Great Lakes system.

The Commission concludes in its report that, based on information and advice from a variety of sources, persistent toxic substances are too dangerous to the biosphere and to humans to permit their release in *any* quantity. When results of the many studies that indicate injury or the likelihood of injury to species throughout the food chain are considered together, it finds that the weight of evidence is sufficient to reach this conclusion.

As a result, the Commission suggests several steps for action. While it recommends that Governments review but not renegotiate the Agreement (as required after every third biennial report from the Commission), it does recommend that the Agreement definition of a persistent toxic substance be revised. This revision would include "those substances with a half-life in any medium — water, air, sediment,

Courtesy of Focus on International Joint Commission Activities soil or biota — of greater than eight weeks, as well as those toxic substances that bioaccumulate in the tissue of living organisms." The present definition refers only to those substances with a half-life in water of greater than eight weeks. Half-life is that time required for a substance's concentration to diminish to one-half its original value.

Because actions to date have not sufficiently reduced or eliminated certain chemicals, including polychlorinated biphenyls (PCBs), DDT, dieldrin, toxaphene, mirex and hexachlorobenzene, the Commission also recommends that these persistent toxic substances be sunset as soon as possible. Sunsetting is a process to restrict, phase out and eventually ban the manufacture, generation, use, transport, discharge and disposal of a substance. Those uses of lead and mercury that result in their discharge or disposal into the environment should also be sunset.

The Commission received a great deal of information and advice over the past two years concerning the use of chlorine in the Great Lakes basin. The Commission recommends that the Parties consult with industry and other affected interests to develop timetables to sunset the use of chlorine and chlorine-containing compounds as industrial feedstocks.

The recently evolving programs created by governments, interest groups, municipalities and industries, the Commission concludes, begin to focus on the specific issues facing the Great Lakes ecosystem and are encouraging signs of action. For example, the Binational Program to Protect the Lake Superior Basin, announced last fall by the Governments of Canada and the United States in cooperation with Michigan, Minnesota, Wisconsin and Ontario (see *Focus*, Volume 16, Issue 3, page 6), includes several provisions to restore and protect the basin through special designations, and pollution prevention and enhanced regulatory programs.

The Commission voices its support for this program and recommends that the governments also establish a deadline for eliminating point or direct releases of persistent toxic substances into Lake Superior or its tributaries. It also suggests that the Parties agree to prohibit new or increased direct discharges and establish a coordinated phaseout of existing sources.

The *Sixth Biennial Report* also includes recommendations on further strategies to sustain the Great Lakes-St. Lawrence ecosystem. Several recommendations previously made in the Commission's *Special Report on Great Lakes Environmental Education* (see *Focus*, Volume 16, Issue 1, page 18) are repeated, including calls for increased emphasis on the Great Lakes and environmental education at all age, grade and subject levels, establishment of a Great Lakes Education Clearinghouse, and support for curriculum development and teacher training programs.

Finally, the Commission recommends that the Parties join with states, provinces and local governments to identify and designate sustainable development areas. Areas of high quality that are being pressured by economic growth would benefit from new community-based programs to ensure that development in these areas is sustainable in economic and environmental terms. In particular, the Commission supports the model program for the Grand Traverse Bay region in Lake Michigan and recommends that the Parties support it as the first of these sustainable growth areas. The resulting program could also provide a model for Areas of Concern to strive for once rehabilitated.

Copies of the Commission's *Sixth Biennial Report on Great Lakes Water Quality* are available from its three offices. Contact the International Joint Commission at:

1250 23rd Street NW,
Suite 100
Washington, DC 20440
(202)736-9000

100 Metcalfe Street
18th floor
Ottawa, ON K1P 5M1
(613)995-2984

100 Ouellette Avenue
Eighth floor
Windsor, ON
N9A 6T3
(519)256-7821

OR

P.O. Box 32869
Detroit, MI
48232
(313)226-2170

Municipal waste, nutrient enrichment, and eutrophication

In 1882, 180 people of every 100 000 in Ontario died of typhoid, cholera, or similar diseases. The reason was simple: towns and cities of the day, without forethought, routinely placed drinking water intakes close to sewer outfalls; in Sarnia, for example, the two were only 45 m apart (Koci and Munchee 1984). Society reacted to the first major environmental crisis in the basin by extending water intakes farther into lakes or upstream in rivers and eventually chlorinating drinking water supplies that had become contaminated with sewage (Steedman 1986). In Toronto, chlorination started in 1910 and resulted in immediate dramatic reductions of typhoid fever, tuberculosis, and infant mortality. The epidemics disappeared and the bacterial problems appeared solved. In fact, they were not. In the late 1940s and early 1950s, studies initiated by the International Joint Commission found bacterial levels in some areas triple those observed in the early part of the century. The earlier action of chlorination had served to mask rather than solve the problem. It took another crisis of a different kind to force a comprehensive cleanup of municipal sewage, which remains far from complete today.

In the 1960s, parts of Lake Erie took on a sickly green hue. The source of this unnatural greenness — a harbinger of ecological calamity — was a bloom of blue-green algae in the open water. At the same time, beaches became covered in green, slimy, rotting masses of filamentous algae called *Cladophora*. In deeper water, the decomposition of algae that had fallen to the lake bottom consumed oxygen, rendering the lake oxygen depleted (anoxic). Mayflies and other species disappeared and were replaced by less desirable species, such as sludge worms. Algae fouled fish spawning shoals. There was a precipitous decline in resident aquatic life, which led to the suggestion that Lake Erie was "dying." Ironically, it was suffering from too much life, in the form of algae, due to nutrient-rich, or eutrophic, conditions.

The process of fertilization that causes high productivity and increased biomass in an aquatic system is called eutrophication. The addition of nutrients, such as phosphorus, nitrogen, and potassium, results in rapid growth of algae, in the same way that lawn fertilizers fuel green growth. In the case of the lower Great Lakes and a number of bays and channels, the aquatic ecosystem as it had been in the past was being replaced by a new one.

Everyone had always assumed that, because of their vastness, only local areas of the lakes could be altered by human activities — not whole lakes. Algal blooms on Lake Erie disabused planners of that comforting theory and, in 1972, led Canada and the United States to sign the first Great Lakes Water Quality Agreement. It sought control of phosphorus loadings through the reduction of phosphorus in laundry detergents and reductions of phosphorus from point sources — that is, identifiable, distinct sources, such as municipal sewage outfalls and industrial effluent pipes.

Loadings at point sources have been dramatically reduced since 1972, primarily through upgrading sewage treatment plants (International Joint Commission 1987*a*). To date, over $9 billion has been spent by the Canadian and U.S. governments, and major loading reductions have been achieved — a major international success.

With many point sources showing improvements, the program was strengthened in 1983 through commitments to curtail inputs from non-point sources, such as agricultural drainage and urban runoff. Inputs from such non-point sources are both technically and socially difficult as well as costly to control, and, as a result, it will take much longer to achieve significant reductions.

Decreasing open lake concentrations of total phosphorus clearly reflect substantial reductions in phosphorus loadings. Lake Erie, the southernmost, warmest, and most productive of the Great Lakes, shows the most significant recovery from eutrophication, with a roughly 60% reduction in phosphorus levels since 1983. Although massive algal blooms no longer occur, oxygen depletion in deep waters remains a concern.

Although the phosphorus concentrations in open waters have improved significantly over the past 20 years, many localized nearshore areas still suffer from nutrient enrichment. Of 43 severely degraded Areas of Concern identified by the International Joint Commission, 32 are characterized by elevated levels of phosphorus, including 13 of 17 Canadian Areas of Concern and 21 are characterized by dissolved oxygen depletion in the water (International Joint Commission 1987*a*). Many inland lakes and tributaries also remain degraded due to nutrient enrichment. This is the case in southwestern Ontario and on the United States side of the Great Lakes

Reproduced from: Government of Canada, 1991, *The State of Canada's Environment*.

basin, where agriculture is intense and where small communities continue to discharge untreated sewage (Colborn et al. 1990).

Although the phosphorus problem has stabilized over the last two decades, the concentration of nitrogen compounds has been increasing in all of the lakes and is itself now the cause of growing concern. The increase in nitrogen compounds in the lower Great Lakes appears to be mainly due to increased fertilizer use, while in the upper Great Lakes, direct atmospheric deposition (e.g., acidic precipitation) appears to be the most significant cause (Barica 1987).

The current shift in the ratio of nitrogen to phosphorus could affect the phytoplankton community and have far-reaching effects on food web dynamics (International Joint Commission 1989). The fear is that if phosphorus controls were relaxed, excess nitrogen now in the system could initiate another round of serious eutrophication (Bennett 1986).

Eutrophication, or phosphorus/nitrogen enrichment, represents a simple stress — response mechanism, where the chemical stress and subsequent biological response can be clearly identified. It was the biological response — rotting piles of algae on the beaches and the disappearance of other aquatic life — that ultimately inspired action. People controlled the stress, and, in turn, the ecosystem.

The 11 "critical pollutants" on the International Joint Commission's "primary track"

Chemical	Production and release	Source
TCDD	Unintentional	Created in the manufacture of herbicides used in agriculture, and range and forest management. Also produced as by-products of combustion of chlorinated additives in fossil fuels and chlorine-containing wastes, through production of pentachlorophenal (PCP), and in pulp and paper production processes that use chlorine bleach. TCDD is the most toxic of 75 congeners (forms) of polychlorinated dibenzodioxins, and TCDF is the most toxic of 135 congeners of polychlorinated dibenzofurans.
TCDF	Unintentional	
Benzo(a)pyrene	Unintentional	Product of incomplete combustion of fossil fuels and wood, including forest fires, grills (charcoal broiling), auto exhausts, and waste incineration. One of a large family of polycyclic aromatic hydrocarbons (PAHs).
DDT and its breakdown products, including DDE	Intentional	Insecticide used heavily for mosquito control in tropical areas.
Dieldrin[a]	Intentional	Insecticide used extensively at one time, especially on fruit.
HCB	Unintentional	By-product of combustion of fuels that contain chlorinated additives, of incineration of wastes that contain chlorinated substances, and of manufacturing processes using chlorine. Found as a contaminant in chlorinated pesticides.
Alkylated lead	Intentional	Used as a fuel additive and in solder, pipes, and paint.
	Unintentional	Released through combustion of leaded fuel, waste, and cigarettes, and from pipes, cans, and paint chips.
Mirex[b]	Intentional	Fire retardant and pesticide used to control fire ants. Breaks down to more toxic form, photomirex, in presence of sunlight. Present sources are residuals from manufacturing sites, spills, and disposal in landfills.
Mercury	Intentional	Used in metallurgy.
	Unintentional	By-product of chlor-alkali, paint, and electrical equipment manufacturing processes. Also occurs naturally in soils and sediments. Releases into the aquatic environment may be accelerated by sulphate deposition (i.e., acidic deposition).
PCBs[c]	Intentional	Insulating fluids used in electrical capacitors and transformers and in the produciton of hydraulic fluids, lubricants, and inks. Was previously used as a vehicle for pesticide dispersal. PCBs comprise a family of 209 forms of varying toxicity.
	Unintentional	Primarily released to the environment through leakage, spills, and waste storage disposal.
Toxaphene[?]	Intentional	Insecticide used on cotton. Substitute for DDT.

[a] Use restricted in the United States and Canada.
[b] Banned for use in the United States and Canada.
[c] Manufacture and new uses prohibited in the United States and Canada.
Source: Colborn et al. (1990).

FIGURE 18.11

Status, trends, and signs of toxic contamination in the Great Lakes and connecting water bodies

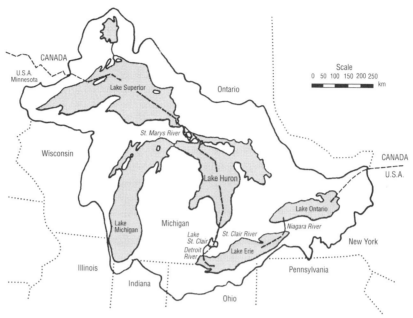

Lake Superior

- best overall water quality in the system.
- 90% of contaminants, including chlorinated organic compounds, enter the lake via the atmosphere.
- lowest metal concentrations; lead, mercury, and cadmium levels elevated in some bays and harbours due to discharges from pulp and paper mills, municipal sewage treatment plants, and mining.

St. Marys River (Lake Superior to Lake Huron)

- least degraded of connecting water bodies.
- localized degradation adjacent to and downstream from steel and paper mills and municipal sewage treatment plants on Canadian side.
- mercury in large specimens of some sport fish exceeds objectives of the International Joint Commission.

Lake Huron

- water quality in open waters almost the same as in the early 1800s.
- contamination in harbours and embayments due to local industry, agricultural runoff, and recreational boating in Severn Bay.
- Serpent River drainage basin contaminated by radionuclides and heavy metals from Elliot Lake mining district.
- concentrations of PCBs, dieldrin, and mercury in trout stable since 1985; DDE, HCB, and PCB levels in Herring Gull eggs reduced by approximately half since 1980.

St. Clair River (Lake Huron to Lake St. Clair)

- major improvements in water quality achieved in the 1980s.
- toxic industrial solvents and metals from Sarnia's chemical valley compromise local environmental quality; HCB and perchloroethylene exceed guidelines.
- sediments, containing toxic organic compounds and mercury from chemical valley, lethal to a number of indicator organisms.

Lake St. Clair (St. Clair River to Detroit River)

- water, sediment, and biota quality improved during the 1980s.
- mercury in sediments a concern, though overall level of metals low.
- toxic organic compounds in fish and duck flesh at potentially significant levels.

Detroit River (Lake St. Clair to Lake Erie)

- an International Joint Commission Area of Concern.
- water and sediments degraded in terms of conventional pollutants, toxic organic compounds, and metals, though improvements achieved since the 1970s.
- a range of contaminants in fish, waterfowl, and other wildlife species.
- oral, dermal and liver tumors in fish in the lower Detroit River.

Lake Michigan

- second only to Lake Ontario in overall degradation by toxic contaminants.
- especially degraded by toxic contaminants in south; sediments have highest levels of PAHs and PCBs in the basin; Waukegan Harbour sediments contain 500 000 ppm (50%) PCBs.
- most severe metal contamination (lead, mercury, cadmium, and others) of all the lakes.

Lake Erie

- heavily polluted by lead, mercury, zinc, cadmium, and arsenic, and toxic contamination close to Lake Michigan.
- metals and organic compounds in both open waters and sediments declining due to burial and dilution.
- Detroit and Maumee rivers major sources of organic chemical contaminants, including PCBs; western basin severely degraded.
- contaminant levels lower than expected in fish species, perhaps due to eutrophic state of lake; in 1987, concentrations of PCBs in spottail shiners in western Lake Erie exceeded objectives.

Niagara River (Lake Erie to Lake Ontario)

- an International Joint Commission Area of Concern; main source of toxins for Lake Ontario.
- 261 human-made chemicals in the system; main compounds of concern are DDT, PCBs, mirex, and mercury, as well as dioxins and furans.
- 57 chemicals considered to pose human health or environmental risk.

Lake Ontario

- most contaminated lake vis-à-vis diversity and concentrations of organic compounds.
- highest mean concentration in open waters of all chlorobenzenes; sediments contain highest concentrations of dioxins and furans; concentrations of problem toxic substances may be stabilizing at unacceptably high levels.
- bioaccumulation of PCBs, TCDD, chlordane, dieldrin and DDT and metabolites in fish makes them unfit for consumption by other wildlife.
- deformities and reproductive failures in fish-eating birds attributed to toxic contaminants.

Source: Data principally from International Joint Commission (1989) and Upper Great Lakes Connecting Channels Study, Management Committee (1988); modified from an unpublished draft figure, prepared by P. Bircham, Environment Canada, State of the Environment Reporting.

Questions and Concerns

READINGS 7 & 8

1. Why would Beluga whales show such concentrations of toxic chemicals in their bodies?

St. Lawrence water deplorable, study says

BY ANDRÉ PICARD
Quebec Bureau

MONTREAL — Forty-four of 45 water-filtration plants along the St. Lawrence River draw water of "deplorable bacteriological quality" because raw sewage continues to be dumped into the waterway, a new federal study says.

This does not mean drinking water is unhealthy, federal scientists stressed, but the high levels of fecal coliform bacteria add tremendously to treatment costs and increase the health risks of workers at the water plants.

"It's a sewer," Marie Lavigne, project director at the St. Lawrence Centre, said of the river, noting that 98 per cent of samples contain fecal coliform. "But the water is 100-percent treated before it comes out of the tap."

One-third of Quebec municipalities continue to dump their sewage directly into the river. At the same time, 43 per cent of the province's residents draw their drinking water from the St. Lawrence.

Bruce Walker, president of STOP Inc., an environmental group that focuses on water quality in the Montreal area, said the public should not be too alarmed by the data. "I'm not prepared to say, based on this data, that all drinking water intakes are at risk."

Mr. Walker said it is obvious that cleaning up the river is the priority, but the public should be far more concerned about the quality of the water that comes out of their taps today than about the quality of untreated water water six years ago.

The 100-page report released yesterday in Montreal, entitled Quality of Water for Direct Human Consumption, is a study of water quality between Cornwall, Ont., and Montmagny, Que., between 1978 and 1988.

The most notable data concerned fecal coliform, the bacteria found in feces.

In the samples taken over a decade, more than 70 per cent of St. Lawrence River water contained more than 100 bacteria per 100 millilitres, while more than 20 per cent contained a staggering 1,000 bacteria per 100 millilitres.

Courtesy of The Globe and Mail

BELUGA CONTAMINATION CONFIRMED

A scientific report released May 27 by World Wildlife Fund (Canada) confirms that St. Lawrence beluga whales are heavily contaminated with toxic chemicals. The three-year research project, funded by various government, industry, and environmental organizations, examined carcasses of belugas and other marine mammals washed up on the shores of the St. Lawrence. Among the toxic compounds found in the animals' tissues were heavy metals such as lead and mercury, organochlorines — including mirex, DDT, and PCBs — and the PAH (Polycyclic Aromatic Hydrocarbon) compound benzo-a-pyrene.

PCB levels were so high that the beluga carcasses could be classified as hazardous waste, according to the study. PCBs, as well as DDT and mirex, are thought to affect reproductive rates in animals. In fact, the St. Lawrence whales proved to have a lower reproductive capability than their Arctic relatives; and mammary lesions in some specimens were severe enough to prevent adequate feeding of any young that might have been born.

Benzo-a-pyrene (BaP), a known carcinogen believed to be associated with tumour formations, is released into the environment through the incomplete combustion of organic compounds in processes like wood burning, engine fuel consumption, and aluminum smelting. BaP metabolites were present in eight of the nine belugas upon which DNA analyses were performed; and 11 of 24 whale carcasses that received autopsies had at least one tumour.

Describing the St. Lawrence belugas as "among the most heavily contaminated animals on the planet," WWF spokesman Steven Price decried "the lack of a concerted attack on the toxic problem (which) continues to place the future of the (species) in jeopardy." To assist in the recovery of the endangered whales, WWF is seeking a reduction through the St. Lawrence Action Plan of the toxic chemical input into the St. Lawrence and Great Lakes Basin. The plan's 1993 target of 90 percent reduction has already been postponed until 1994-95. In calling for an increased rate of implementation, Price also urged that the pulp and paper sector be among the industries forced to meet the original target date. At present, that sector is exempted from complying altogether.

B.S.

Courtesy of Nature Alert, July 1992, published by The Canadian Nature Federation

Groups Demand Freeze on Kodak's Toxic Discharges

by Ed Cooper, Atlantic States Legal Foundation and Diane Heminway, Citizens Environmental Coalition

A freeze on all toxic discharges by Kodak is what Atlantic States Legal Foundation, Citizens' Environmental Coalition and Great Lakes United have asked for from the NY Department of Environmental Conservation (DEC). We also want DEC to mandate reductions in toxic discharges from Kodak's Rochester, New York facility, and to require that Kodak eliminate its discharge and use of persistent toxic substances within five years.

"The draft permit is in direct contradiction to the goals of the Clean Water Act and the Great Lakes Water Quality Agreement," said Ed Cooper, legal counsel for Atlantic States Legal Foundation. "We should be moving towards reductions in discharges; this permit would authorize the discharge of more chemicals to the Genesee River."

Kodak's Rochester facility is one of the largest polluters in the Great Lakes Basin. The Citizen's Fund and the Industrial States Policy Center, in a report entitled "Poisoning the Great Lakes: Manufacturers Toxic Chemical Releases" (1992), identified the facility as the second largest source of toxic chemical discharges (17 million pounds) into the Great Lakes Basin in 1989. A very significant quantity of toxic chemicals, more than 415,000 pounds, were released to water (the Genesee River). Kodak Park was ranked fourth as an industrial discharger of toxics to water in the Great Lakes states (and by far the largest in New York). It was also identified as discharging more known or suspected carcinogens than any other industrial facility in the Great Lakes states, and far more known or suspected carcinogens directly to water than any other facility.

Things have gotten worse since 1989. Kodak's Rochester facility currently discharges as much as 640,000 pounds of toxic pollutants into the Genesee River each year. The Citizen's Fund and the Industrial States Policy Center estimate that the total quantity of known, suspected or experimental carcinogens currently discharged by the facility may amount to as much as 189,000 pounds per year. They also estimate that the facility presently releases as much as 219,000 pounds of known or suspected reproductive toxins and 248,000 pounds of known or suspected teratogens. The draft permit would allow these amounts to increase dramatically.

"The impacts of persistent toxic substances on the Great Lakes-St. Lawrence River ecosystem have been virtually ignored by DEC in the development of this draft permit," said Karen Murphy, a GLU field coordinator. "DEC should require Kodak to eliminate the use and discharge of persistent toxic substances altogether."

The groups recommended that the DEC amend the permit as follows:
- Set limits for all pollutants authorized to be discharged.
- Impose an initial freeze on current discharges. The limits should then be made more stringent in each successive year to ensure that Kodak's toxic discharges are reduced significantly over a five-year period.
- Require Kodak to make a serious advance towards achieving the goal of zero discharge of persistent toxic pollutants. By the final year of the permit, the discharge of these pollutants should no longer be authorized.
- Develop timetables and programs for sunsetting the use of chlorine and chlorine-containing compounds.
- Require regular and frequent monitoring for all pollutants authorized to be discharged, as well as, for a number of highly toxic substances that are manufactured and used in large quantities at Kodak's facility.

The current permit, issued in 1984, was supposed to be in effect for five years. It has taken DEC and Kodak three years to negotiate the new permit, while Kodak continues to operate under the 1984 permit, thus making it, in essence, an eight-year permit.

It is unacceptable that in 1992 companies are still allowed to discharge hundreds of pounds of chemicals each day that are known or suspected of causing cancer and birth defects. The DEC should be banning poisons in the public's drinking water rather than issuing permits for their discharge.

Courtesy of Great Lakes United (Fall/Winter 1993)

Discussion for Toxic Pollution in the Great Lakes

1. The Great Lakes are shared by Canada and the United States. Environmental cleanup challenges are often complicated by this type of relationship if one country is a greater polluter than the other (Russia and Canada in the Arctic is an example; Mexico and the United States at their common border is another). From the evidence in these readings, which country – Canada or the United States – poses the greater threat to the lakes given domestic environmental legislation and the pollution flowing from cities and industries? Which country is doing more to protect the lakes?

2. If the St. Lawrence River is polluted with toxic chemicals, why do you think officials and environmentalists are unwilling to say that drinking water supplies are affected? This calls into question the concept of "acceptable risk," where scientists determine what amounts of certain chemicals can be ingested by humans before the risk becomes unacceptable. Is this a practical concept? What might be wrong with its basic premise? Why is there a concept of "acceptable risk"?

3. Should companies be allowed to pollute the common water resources we all share? Why do you think corporations are still legally allowed to dump toxic chemicals into Great Lakes waterways while animals living in the basin are shown to have reproductive abnormalities and people downstream drink the same water? If a company says it cannot comply with "zero discharge" regulations, should it still be allowed to operate? Consider an "economy vs. environment scenario," should jobs matter more than the environment? Who should decide this?

Introduction

Forestry

This issue examines the forest industry in Canada. It looks at the relationships between government and forestry companies; how trees are harvested and some possible alternatives to clearcutting – the most destructive form of harvesting; and the importance of preserving biodiversity within Canada's forests.

Canada's forests are one of its greatest natural assets. Almost half the country's land mass is forested or capable of supporting forests, and it possesses the world's single largest national inventory of softwood forest. The value of Canadian forests, based on the contribution of the forestry sector to Canada's gross domestic product, is $20 billion per year. The forest sector employs around 900,000 Canadians who earn roughly $9.6 billion annually. Its non-monetary value is incalculable. Canada's forests are the habitat for much of the country's wildlife and insect species; a rich breeding ground for a diversity of plant life; the hunting ground for thousands of aboriginal people; an important recreational resource; and a vast natural ecosystem that purifies air and water, conserves water, regulates climate, and absorbs carbon on a global scale.

Like Canada's fisheries, its forests are harvested aggressively. The method-of-choice, used by major forest companies, is clearcutting, where vast tracts of trees are removed with chainsaws and large machines called "feller forwarders" that shear, strip, and stack tree after tree in one mechanized process. While clearcutting may be economically efficient, it destroys forest ecosystems and greatly impairs a forest's ability to regenerate itself. In a clearcut, the most opportunistic tree species, such as aspen and willow, tend to overwhelm the seedlings of more valuable softwood species such as spruce, pine, and tamarack, changing the "genetic personality" of the forest. Animals that were once a critical part of the ecosystem, aiding its growth through the spread of fungus and seeds, are unable to survive in a clearcut.

Clearcutting is not inherently bad. The larger problem in Canada has been improper forest management by companies determined to extract the maximum allowable timber, with the blessing of governments in need of the resulting revenue. Clearcuts have historically been too large (the environmental group Forests For Tomorrow found one 200,000-hectare clearcut near Kapuskasing Ontario); replanting and follow-up silviculture such as weeding and fertilizing has not been adequate; enormous tracts of forest have been consigned to future commercial exploitation without proper public consultation; and forest companies have refused to adopt or, in some cases, even consider more sustainable methods of timber extraction.

The Canadian federal government maintains an approximate inventory of forested lands in Canada, but no one can estimate accurately what has been cut, what has not been cut, and what remains productive. Based on government statistics, about 65% of Canada's boreal forest (the predominate northern belt of mostly softwoods that stretches from coast to coast) has been logged at least once. The remaining 35% is a fragmented collection of virgin forest, much of it commercially unviable.

Behind these estimates there are ominous signs that the resource is in serious decline. Companies cut more than twice as much timber today as they did in 1950 and clear almost twice as much land. Of that, over 44% fails to regenerate into productive forest – an area about the size of Prince Edward Island. More than 10% of all Canada's productive forestland has been ruined for future timber extraction. And although tree planting has increased since 1980, it has not yet offset the growing backlog of non-productive clearcuts.

On the bright side, it is estimated by the federal government that over 80% of newly cutover forests are regenerating properly, either on their own or through reforestation programs. This is an increase of 22% from the early 1970s. Federal and provincial spending on silviculture rose by more than fourfold between 1979 and 1988, from under $50 million to more than $200 million.

But some provinces remain negligent. Quebec has a particularly sorry record. Of the forests clearcut in Quebec between 1980 and 1988, only 39% have been reforested naturally or artificially; only 7% of areas harvested since 1985 have been weeded, thinned, or fertilized. And, as with all other industry reforestation programs, single-species monocultures are typically planted in place of the forests that are removed.

Foresters and environmentalists in Canada and the United States, frustrated with simply complaining about the situation, are suggesting alternatives to destructive logging and monoculture reforestation. The most talked about alternative management strategy is "New Forestry," an approach that might also be called "selective extraction." Still evolving from research and yet to be practiced on a large scale, New Forestry is based on the premise that if forests are to survive they should be treated as ecosystems not fibre factories. Its proponents call for the protection of forest soils and the preservation of biodiversity by logging in ways that imitate natural patterns: leaving wood and debris to rot on the forest floor; making smaller clearcuts, perhaps in narrow strips so living forest adjacent to the clearcut portion will influence regeneration; and allowing some trees to remain standing to spread seeds, provide shade, and harbour wildlife.

Others advocate a new approach to land ownership. In Canada, as in the United States, much destructive clearcutting occurs on public land leased or licensed to forest companies at very low prices (often so low it is debatable what benefit is being derived by the public that owns the land). In Sweden, 75% of the forests are privately owned by small landowners who choose what to cut and when to cut it, thereby ensuring the land will yield timber sustainably. A variation of this is the community forest – tracts of forested lands owned and harvested by small communities for their own benefit. In British Columbia, where two communities – North Cowichan on Vancouver Island and Mission in the Lower Mainland – have owned forests since the 1930s, the lands have been a steady source of revenue, recreation, and education while remaining commercially productive.

Unfortunately, forest overexploitation in Canada could remain tragically massive for many years into the future because of deals made between provincial governments and private companies. As Christie McLaren writes in *Equinox*: "Altogether, 45 developments – 24 new or expanded pulp mills, 7 paper mills, 9 sawmills, 2 simulated-plywood plants, and 3 chopstick factories – worth more than $10 billion are under construction or consideration across the northern forest, from British Columbia to Quebec. With little or no public consultation, provincial governments have opened public lands, and frequently the public purse, to multinationals from North America, Korea, Hong Kong, and Japan." In Alberta, the government leased 15% of the province's entire land base to two Japanese companies. Manitoba has granted one company – Repap Enterprises Ltd. – the right to log one-fifth of the province or 77% of its prime forested lands. Almost 100% of Canada's most productive boreal forest is now committed to 20-year leases and thus open to logging, including several provincial parks and wildlife reserves.

The issue in Canadian forest harvesting, as with fishing, is one of restraint, or lack of restraint. A renewable resource like trees or fish will only renew itself if given the opportunity. Overharvesting and habitat destruction will ruin it. Sustainable forestry, in tune with biological systems, is the only way to guarantee Canada's magnificent forests remain a valuable public heritage.

Questions and Concerns

READINGS 1 & 2

1. Cree and other aboriginal peoples are sometimes called "forest dwellers" because they live and hunt in the northern forests of Canada. If 100% of the country's boreal forests are committed to forestry leases, where does this leave the people who have traditionally lived off those forests? Should native rights be considered over economic opportunity when it comes to timber extraction and pulp making?

2. The long, strong fibres of northern softwoods like black spruce and white spruce have helped Canada produce some of the world's finest pulp and paper. Now that these species are overexploited and less accessible to pulp companies, a growing number of mills are adopting new technologies for converting aspen – a fast-growing hardwood that thrives in many clearcuts – into marketable paper. Consider the environmental implications of this strategy.

3. In granting long leases on public lands to forest companies, provincial governments seem to be sacrificing a public heritage for relatively little revenue. Why do you think this is happening?

4. As you read these two articles, keep a running list of the negative effects of large clearcuts.

HEARTWOOD

THE WHOLESALE SELL-OFF OF CANADA'S BOREAL FOREST IS NEARLY COMPLETE. CANADIANS ARE FINALLY BEGINNING TO TAKE NOTICE.

Article by Christie McLaren

Heat and steam and the locker-room smell of chlorine envelop Alan Twa like a blanket as he steps from the cool corridor into the raucous heart of Saskatchewan's only pulp-and-paper mill. Clamping on headphones against the din, the short, compact technical manager for Weyerhaeuser Canada Ltd., Saskatchewan Division, moves across the concrete with purpose, into the plant's primary pulping division – a spaghetti network of tubes, boilers, gauges and switches – past a pressure cooker as tall as a 20-storey building, through air that reeks of ripe fish, skirting vats of inky "black liquor," down stairs and up stairs, all the while avoiding the pungent yellow goo that drips on the floor from overhead pipes and shouting explanations above the roar of machinery.

Twa stops before a row of giant metal drums suspended above the floor. Here, dirt-brown wood pulp the consistency of chewed celery is passed through a series of six chemical baths and emerges, nine hours later, white. Climbing onto a catwalk, he reaches inside the last dangerously spinning cylinder and pulls out a clump of wet pulp, white as snow. "See how bright that is?" Despite years of experience, he seems as delighted as a child with this fresh discovery. He pauses a moment, admiring the pulp, before adding, "Some folks say it's too bright." When asked who, Twa manages a wry little smile. "Greenpeace," he replies.

Every day, Weyerhaeuser's giant Prince Albert Pulp and Paper Complex turns the equivalent of 30 football fields of Boreal forest into pulp. And every day, it spews $2\frac{1}{2}$ tonnes of chlorinated compounds, including minute amounts of toxic dioxins, into the North Saskatchewan River. The company has just added a $250 million paper mill to its 1,000-tonne-a-day pulp mill, and officials say they are considering a 60 percent expansion in logging, if markets permit. To Twa and his fellow employees, this remarkable consumption of trees heralds the best and biggest boom in pulp and paper that the Prairies have ever seen.

To others, however, including a rapidly growing number of ordinary Canadians as well as thousands of forest-dwelling aboriginals and a host of noisy environmental groups, it is cause for alarm. The result has been a heated debate over the monopolization of public lands, the destruction of wild Boreal forests and the poisoning of northern rivers.

Altogether, 45 developments – 24 new or expanded pulp mills, 7 paper mills, 9 sawmills, 2 simulated-plywood plants and 3 chopstick factories – worth more than $10 billion are under construction or under consideration across the northern forest, from British Columbia to Quebec. With little or no public consultation, provincial governments have opened public lands, and frequently the public purse, to multinationals from North America, Korea, Hong Kong and Japan.

Some of the schemes give new meaning to the word "megaproject." While the Alberta government has leased 15 percent of the province's land base to two Japanese multinationals, the government of Manitoba has granted one firm, Repap Enterprises Ltd., the rights to nearly one-fifth of that province, or 77 percent of its prime forested lands. In total, Crown forests several times the size of Great Britain, including hitherto untouched stands of timber, have been consigned to a future as facsimile paper, toilet tissue, newsprint, glossy magazines and chopsticks. Nearly 100 percent of Canada's most productive Boreal forest, including several provincial parks and wildlife reserves, is now locked up in 20-year leases and available for logging.

For the Canadian Pulp and Paper Association, the new developments are proof that Canada is holding its own in the fiercely competitive global market-place. To a country that already produces 10 percent of the world's timber,

Courtesy of Christie McLaren

"zero growth is stagnation," says Louis Fortier, the veteran spokesperson for the industry association. "This is true of a company, and it's true of an industry, and it's true of world demand. Expansions are necessary. If we don't do it, somebody else will."

Environmentalists and scientists alike, however, are wondering whether Canada's remaining Boreal forest can withstand such expansion without critically weakening the Boreal ecosystems, which sustain life in more than half the country. Social critics, armed with the results of recent research, new theories and suggested new practices, claim that the provincial governments are jeopardizing the continued well-being of public and provincial Crown lands to create social benefits in the form of jobs and tax revenues. The governments have done this by encouraging the pulp-and-paper industry to monopolize the resource, through millions of dollars in government subsidies. In spite of government assurances, they argue, the new generation of forest-management agreements continues to emphasize the production of timber over the protection of the land, wildlife and other values of the forest. The result is that Canada is squandering what remains of its natural heritage by exporting raw pulp to Japan, Korea and Taiwan.

None of those watching the debate, however, have more at stake than the tens of thousands of Cree who know the forest as home. For many of them, new roads and mills could mean an end to the only way of life they have ever known. Clear-cutting and hydroelectric dams have already undermined much of the "forest dweller" culture in Canada. Similar plans for the future, they say, could destroy it. "Native people are fighting for their lives," says Lorraine Sinclair, a 37-year-old Metis from Edmonton who is executive director of the Mother Earth Healing Society. "Our culture depends on the forest being there. When we are talking about the 'great Boreal sell-off,' we are talking about cultural genocide."

The difference between those who live in the forest and those who want to exploit it for profit is starkly etched in everyday language. While government and industry officials reduce the forest to its commercial parts, speaking of "fibre" or "timber" or "cubic units," the Cree have more than 200 words for the forest, each a complete thought. *Ehmistikaskat* is "a place in the state of being a forest," while *ehmichumskat* is "the place that produces a lot of food." In their tongue, trees are living beings and often a source of spiritual power. The heartwood of a tree is "the child wood" or "the son wood." Cree tales refer to "the standing-tree person." And while government and industry speak religiously of global markets and jobs, the Cree invoke the name of *Chuetenshu,* the North Wind and Lord of the Animals who cares for the forest.

Russell Willier, a cheerful 40-year-old medicine man from the Sucker Creek Reserve in northern Alberta, says that the Cree look for "a forest that hasn't been touched" when it is time to fast and pray. "The spirit world is closer there," he explains, "than it is where it is all raped down." But untouched forest is getting harder to find. Willier's trapline in the nearby Swan Hills was reduced to stumps last year, and his ancestral moose-hunting country south of the reserve has been cleared by homesteaders.

Exactly how much primary Boreal forest is still standing in Canada is anybody's guess; governments have not kept track. The best estimate, based on federal statistics for all commercially productive forests, is that 65 percent has been logged at least once, leaving virgin fragments that total just 35 percent – an area half the size of British Columbia. As in the Tropics, the loss of mature and old-growth forests means the disappearance of an important natural pharmacy, known to aboriginal peoples but barely tapped by Western medicine. Willier has gained international recognition for treating stubborn cases of psoriasis with remedies concocted from such Boreal plants as aspen and chokecherry. But as the forests shrink, so does Willier's pharmacy.

The forest the Cree call "home away from home" is part of the largest natural ecological zone on Earth. Its lonely beauty, the product of a haunting union of rock and tree and water – where the mournful cry of the loon blends with the sociable howl of the wolf – has inspired art and literature in both English and French Canada and is the foundation of native religion and culture. From the southern Yukon to the northeastern tip of Newfoundland, some three million square kilometres of black spruce, white spruce, jack pine, birch, tamarack, fir, balsam poplar and aspen soften the bony spine of the Canadian Shield with a spiky tapestry of green. Flecked with lakes and wetlands, this sub-Arctic taiga is home to a wide variety of plants and animals and nurtures its own subspecies of Boreal lillies and orchids, toads, frogs, bats and butterflies.

In Alberta alone, some 100 plant species are found only in the Boreal forest. And Peter Lee, manager of natural and protected areas for the provincial government, says as many as half of these are rare, including the yellow Indian tansy and the cypripedium orchid, a bloom as big as a baby's fist. The forest is also the only habitat of the woodland caribou, the Baird owl, the Cape May warbler and dozens of other songbirds that spend their summers in Canada and their winters in the Amazon. The Boreal forest conserves the soil, regulates the flow of water on a vast piece of North America and, say scientists, is second only to tropical forests in its capacity to store carbon as a protection against global warming.

The forest renews itself in an unbroken cycle of birth and death through complex mechanisms that science has only recently begun to appreciate. Ruled by bitterly cold winters and short growing seasons, a stand of black spruce can take 150 years to make the transition from grasses to fast-growing hardwoods to a dense tangle of diverse vegetation to the mature growth the Cree call "the forest you can see through." When old and infirm, trees are felled by fire, insects or wind, opening the way once again for grasses. Although dead and decaying trees have no commercial value, they are a source of nutrients and water that are crucial to healthy reforestation, and every organism, from the microscopic spore to the moose, has a role to play. So intricate is the web of interactions that the droppings of just one deer mouse can contain enough spores of mycorrhizal fungi, nitrogen-fixing bacteria and yeast to inoculate several black spruce seedlings. "A forest is like a cake mix," explains Chris Maser, a leading forest biologist formerly with the U.S. Bureau of Land Management. "The ingredients are half the cake, and the interactions among the ingredients are the other half. It's more complicated than a NASA space shot."

The forest is so complex, in fact, that scientists do not know how to re-create it once it is cut down. "The Boreal forest is the least studied of any forest type in North America," says biologist Lee. "We know only a few of the parts of existing forests, so how can we even pretend to know the whole?" Foresters, who may be masters at growing some kinds of trees, are in the dark when it comes to the thousands of ingredients and invisible interconnections that enable the Boreal forest to function as a resilient organism. To add to the confusion, the Boreal region consists of a dozen or more distinct forest types, each one a unique mix of species and conditions. We are ignorant, in most cases, about how to reproduce mixed-species stands; how fungi and microorganisms interact with animals and trees; which songbirds eat which insects; and which plants, if any, are weeds.

It also seems that neither industry nor government is eager to learn the answers to such ecological riddles. George Marek, a retired forester who served with the Ontario Ministry of Natural Resources, say that when forest ecosystem research began 20 years ago in the United States, Canada missed a golden opportunity "due to the reluctance of government and industry." Scientists who pioneered ecosystem research at Yale University approached Marek in the early 1970s with a proposal to document his strip-cutting methods of regenerating black spruce near Nipigon, Ontario. But the province and industry had so many objections to the project, says Marek, including worries about financing and the publishing of statistics about forest growth, that the scientists eventually gave up. "It was just frustrating," he says, "because we would have benefited tremendously."

At least part of the possible cost of that missed opportunity can be glimpsed in recent reforestation statistics. Today, companies annually cut more than twice as much timber as they did in 1950 and strip almost twice as much land of trees. In the past decade alone, the Canadian industry has felled a forest the size of East Germany. Yet according to an analysis of statistics published this year by Forestry Canada and the Canadian Pulp and Paper Association, almost half of it – 44.5 percent – has failed to regenerate (the cutting stripped off soils and mosses and left bare rock) or is of "unproven stocking." This means that the status of the cut land is unknown, or it is not growing a forest of sufficient density or quality to support an industry in the future.

At this rate, of all the forest cut in Canada each year, an area the size of Prince Edward Island fails to reestablish itself and may *never* reestablish itself. Since the assault on North America's forests began over a century ago, more than 10 percent of Canada's productive forestland has been so devastated that it is unable to reproduce any sort of merchantable timber at all. The nationwide increase in tree planting during the past decade has yet to make a dent in the growing backlog of overexploited land. According to Gordon Baskerville, the dean of forestry at the University of New Brunswick, Canada has become the "slum landlord" of a degraded public forest.

Canadian taxpayers are also becoming poorer landlords. In a 1989 study of timber supply for the Canadian Forestry Service, Woodbridge, Reed and Associates, consultants for the forestry-products industry, warned that "the timber harvest cannot continue to grow at historic levels." In short, much of what Canada has lost cannot be returned.

Lumber manufacturers have temporarily survived the diminishment of Canada's forests because of new technology that puts smaller trees and pieces of trees to profitable use. Solid two-by-fours are giving way to laminated wood. Plywood has given way to chipboard, particleboard, Aspenite and now something called "oriented-strand board" – a marvel of engineering that blasts trees apart to isolate their fibres, then glues them back together again in a way which mimics the strength of the original. Michael Fitzsimmons, a Saskatchewan environmentalist from Prince Albert, says simply, "They look at it as a technological achievement, whereas I look at it as a sign that the resources are being depleted."

Among the species in short supply are the slow-growing, top-grade black spruce and white spruce, trees whose long cellulose fibres have allowed Canada to become known as a producer of the finest pulp and paper in the world. Across the country, however, companies are now running out of spruce and other softwoods within an economical distance of their mills. And to their chagrin, large-scale clear-cutting – a logging method that was once believed to mimic natural fires and to replicate the Boreal forest – often produces a very different forest from the original, dominated by a single species, aspen, the tree the Cree call "noisy leaf."

The new stands of aspen – a skinny hardwood whose root suckers shoot up like a weed after logging or fire – could have proved disastrous to the forest industry. But with new technology and a little creativity, researchers discovered that the creamy flesh of the aspen will produce a bright white pulp and a fine smooth-surfaced paper which resists yellowing. Accordingly, a whole new market was opened, and now, because of the world's growing demand for paper, it is estimated that Canada's production of bleached hardwood pulp will have increased by 36 percent by 1991. Aspen's newfound popularity, though, is not the only development fuelling pulp-mill expansions in the Boreal forests. The primary drive is cheap trees, unknowingly subsidized by provincial taxpayers.

Saskatchewan provides a salient example. In 1986, Weyerhaeuser Canada Ltd. bid for and won the Saskatchewan government's Prince Albert Pulp Co. Ltd., a nearby sawmill and a lease for 34,000 square kilometres of Boreal forestland. In granting the company a 20-year Forest Management Licence Agreement, Saskatchewan

A Canadian/Amazonian Index

CANADA'S TRANSFORMATION OF THE BOREAL FOREST RIVALS Brazil's exploitation of the Amazon. In search of economic development and additional jobs, each country is providing developers with subsidies to flood or log their most important watershed. In both countries, economic forces have also conspired to clear forests for farmland, treat rivers and oceans like sewers, poison fish and drive native people out of their ancestral homes.

• Size of Canada: 9.9 million square kilometres
• Size of Brazil: 8.5 million square kilometres

• Size of all Canadian forestlands: 4.5 million square kilometres (45 percent of country)
• Size of Canadian Boreal forest region: 3.3 million square kilometres (34 percent of country)
• Size of Brazil's Amazon rainforest: 3.5 million square kilometres (41 percent of country)

• Amount of all Canadian forests cleared each year: 12,220 square kilometres
• Amount of Brazilian Amazon cleared or burned each year: 35,000 square kilometres

• Average amount of timber produced by Canada annually from 1984 to 1986: 421 million cubic metres
• Average amount of timber produced by Brazil annually; 1984 to 1986: 493 million cubic metres

• Estimated amount of Canadian forest that is regenerating to a productive new forest (capable of supporting an industry in the future) within five years of logging: 55 percent
• Amount of Brazilian Amazon that regenerates to a productive new forest after logging: virtually none

• Amount of productive Canadian forest that is now barren or "not sufficiently restocked" with a quality or species of tree capable of continuing to support industry: 10.3 percent, or 250,000 square kilometres
• Amount of Brazilian Amazon that has disappeared: 12 percent, or 420,000 square kilometres

• Estimated number of species in Canadian Boreal forests: 25,600 to 27,600 (58 mammals; 200 birds; 79 reptiles and amphibians; 22,000 insects; 50 trees; 1,200-2,200 flowering plants; 2,000-3,000 fungi)
• Estimated number of species in Brazilian Amazon: between 1 and 2 million (known species: 125 mammals; 400 birds; 100 reptiles; 60 amphibians; 3,000 fish; 300,000 insects; 750 trees; 55,000 flowering plants; 50,000 fungi)

• Amount of Canada's Boreal forest region protected from development and commercial extraction of resources: 2.6 percent, or 85,318 square kilometres
• Amount of the Brazilian Amazon protected (in parks, research sites and extractive reserves): 9.4 percent, or 328,704 square kilometres

• Estimated number of Indians and Metis living in the Boreal forest: 100,000
• Estimated number of Indians living in the Amazon forest region: 170,000

–Christie McLaren

offered Weyerhaeuser a loan guarantee worth about $83 million and a pledge to build 32 kilometres of roads annually. In return, the company promised to construct a 235-job paper mill and to modify the pulp mill to use much more aspen. On paper, the deal was worth $236 million to Saskatchewan taxpayers, but Weyerhaeuser did not pay a penny. Instead, Saskatchewan accepted a 20-year debenture for the full amount with an interest rate of 8.5 percent. The government expects to see the money within 30 years (there is no firm repayment schedule), but that will happen only if the mill makes a 20 percent return on the investment – a feat which would place it far ahead of most mills in North America. The province has, however, at least reaped a $60 million "windfall" under a clause that obliged the company to forward the bulk of any profits it made in the first year of the deal.

In addition to the government subsidies, Weyerhaeuser further benefits by paying some of the lowest timber royalties in Canada. It pays the government 50 cents to $1.25 for a cubic metre of spruce and 31 cents for aspen. Even after contributing an additional $2.30 per cubic metre of softwood and 50 cents for hardwood to a required reforestation fund, the company – using two-thirds softwood and one-third aspen – pays about $14.50 in royalties to produce a tonne of market pulp worth $850. The company's plan to use more aspen is shrewd; the same tonne of pulp from 100 percent aspen would cost just $5.31.

Similar deals, all made secretly, have taken place across the country. Alberta, for instance, which charges even less for its timber than Saskatchewan does, has subsidized its controversial pulp-mill expansions with more than $1 billion in loan guarantees, roads and railroads. In Manitoba, the government turned over an ailing Crown forestry corporation to the Montreal company Repap Enterprises Ltd., along with $240 million in loan guarantees and timber rights to 17 percent of the province. In exchange, Manitoba got $10 million in cash and $122 million in preferred shares.

Besides such secrecy, plans to expand logging have been characterized by a paucity of research about their effects. In Saskatchewan, for example, Weyerhaeuser says that its aspen expansions are sustainable. But Saskatchewan's top government forester cannot confirm that claim with absolute certainty. Timber licences in Saskatchewan are renewed every five years if the company satisfies minimum government reforestation standards. But Murray Little, director of forestry for Saskatchewan Parks and Renewable Resources, notes that his staff has been stretched thin by budget cuts and cannot inspect all the plantations they should.

As a consequence, most of the government's and Weyerhaeuser's pronouncements on the state of the forest must be taken solely on faith. Michael Fitzsimmons of Prince Albert says that he has been waiting in vain for two years now for government officials to tell him exactly how much of the public forest leased to Weyerhaeuser has been clear-cut, how much has been burned, how much is naturally regrowing, how much was reforested, how successful it was and how fast the trees are growing. "They have a hard time giving me this information," he says in frustration. "I don't think they have it. I don't think the provincial government knows what's going on."

In reply, Little says that the detailed information about the Crown forest which Fitzsimmons wants is the private property of Weyerhaeuser. "Where the company has spent money and time putting it together, we don't think we'd be responsible if we released it to just anyone," he says. "When people want something very specific, it makes you wonder why they want it."

Corporate secrecy has not stopped the public from wanting specific answers about water pollution. To produce the bright white pulp their customers demand, Weyerhaeuser and dozens of other manufacturers use chlorine-based chemicals, discharging hundreds of thousands of tonnes of toxic waste into rivers and oceans every year. Among them are dioxins and furans, which build up in the food chain and, according to the federal Department of Fisheries and Oceans, "constitute a threat to human health" for people who regularly eat contaminated fish. In response to public outcry, Ottawa has begun to pressure the industry to clean up its effluent. Last year, the government shut down several west-coast shell fisheries because of dioxin contamination from pulp mills, and in April, it proposed new regulations that would virtually eliminate the most dangerous dioxins and furans from mill wastewater. Nationwide, the industry estimates it will cost $5 billion to comply.

But many industry experts, including Weyerhaeuser's Alan Twa, label the demand for expensive antipollution measures "a lot of hype." They say there is no proof that minute amounts of dioxins or other organochlorines have hurt the environment or people. In his spartan office in the Prince Albert mill, Twa stresses that fish tests in the North Saskatchewan River last year did not indicate dangerous levels of 2,3,7,8-TCDD, the most toxic dioxin. "I don't see fish dying," he says with a flash of anger. "The mill's been here for 20 years already, and it hasn't happened yet. The evidence isn't there." In 1988, federal scientists who pulled white suckers from the river one kilometre downstream from the Weyerhaeuser mill found minute – yet acceptable – amounts of the most toxic dioxins and furans.

No one knows, however, what the effect will be on the fish and plants – and the humans who rely on them – as more and more new mills and expanded existing mills add their burden to the northern rivers. The Repap mill expansion, for example, changes its status from a small producer of unbleached paper to the largest bleached-paper producer in the country. When finished, it will add three new tonnes of effluent each day to that already dumped into the North Saskatchewan by the Weyerhaeuser operation. The concern is one of cumulative effect, but at present, there is no control over the total burden a river must bear. Even under Ottawa's proposed new legislation, the amount of pollution permitted from each mill is based on the amount of pulp it produces, not on the river's ability to bear the chemical load. Lyle Lockhart, head of contaminants and toxicology research at the Department of Fisheries and Oceans in Winnipeg, says that such an arrangement "makes absolutely no biological sense. The river doesn't get bigger or smaller with the size of the mill." A bioaccumulation of toxins in aquatic life could affect fish and the natives who eat them, even as far downstream as lake Winnipeg, he speculates.

In Manitoba, the sheer size of Repap's logging expansion – the company will eventually build 2,000 kilometres of roads to service the cutting of a wilderness area the size of East Germany – will also have consequences. It could very will make "environmental refugees" of what is believed to be the highest concentration of aboriginal forest dwellers in North America. Thousands of Cree leave their reserves each spring to live in camps in the forest, returning in late fall with a full harvest of game, fish, berries and herbs. According to Michael Anderson, research director for the Manitoba Keewatinowi Okimakanak, which represents 27,000 Indians on 23 northern reserves, the Cree's dependence on the forest eerily parallels that of Amazonian Indians. Last year, Chief Paulinho Paiakan of Brazil's Kayapo Indians shocked a Canadian friend with a desperate telephone call from Europe. Finding himself in an alien culture, cut off from the familiar edible plants and animals of the rainforest, Paiakan reported that there was nothing to eat and that he was starving. Anderson has received similar calls from Cree living in Winnipeg.

Neither Manitoba nor Saskatchewan is alone in felling forests and uprooting native people. Quebec, the undisputed giant of Canada's pulp-and-paper industry, has a singular record of environmental degradation. Brian Craik, an anthropologist and director of federal relations for the Grand Council of the Crees of Quebec, says that timber harvesting at the southern fringe of the James Bay region alone is eating up 600 square kilometres of forest a year. That is the size of a hunting territory which feeds about 30 people. In the past decade, 12 to 15 such areas have been destroyed, says Craik; the families who depended on them have turned to welfare and are experiencing social difficulties such as alcoholism, violence and family breakdowns. He predicts that the remaining hunting territories in the region will be wiped out within 20 years. the next generation of Indians "will be pushed off the land…and I think experience tells us that they can't go back."

Some people, however, suggest that there are other considerations, like jobs, which must be given their due. André Lafond, co-president of the Ministry of Energy and Resources team responsible for negotiating 300 new forest licences, says that *all* of the accessible Crown timber in Quebec is slated for logging. He confirms that even wildlife reserves such as La Verendrye, which is claimed by the Algonquin Indians, are reserves in name only. "That's what they call it, period," he says. "Closing this area to logging would mean 4,000 to 5,000 people would lose their jobs."

Yet many experts doubt that Quebec's remaining forests can sustain such logging. New mills are being approved "mainly to solve the unemployment situation," says Gilles Frisque, director of a multiregional forest research centre at the University of Quebec. He says he is not as confident as the government that the wood is there. "But when an industry is willing to invest tens of millions of dollars," he says, "the government is ready to give them the timber they need."

Quebec's government dismisses its critics. "They don't know what they are talking about," says Lafond. He explains that every stand of forest licensed to the industry this year has been carefully calculated by computer. However, the most telling parts of those calculations are not available to the public. As in Saskatchewan, Lafond says that it could hurt a company's competitiveness to release information showing how well its raw material is growing on public land.

What few statistics are available, however, indicate that Quebec's record of managing its timber is among the worst in the country. Of the forests that were clear-cut between 1980 and 1988, Quebec claims to have reforested only 39 percent by natural or artificial means. Since 1985, just 7 percent of the harvested area on average has been weeded, thinned or fertilized, compared with a national average of 25 percent. According to analysts Woodbridge, Reed and Associates, the forests are changing in composition: second-generation forests originating from large clear-cuts contain more hardwoods such as birch and poplar, although the industry remains dependent on softwoods. Even Albert Côté, the province's forestry minister, admitted in a brochure last year that Quebec's "forests are in such a pitiful state that we are about to face a shortage in many regions" and, eventually, across the province.

In an effort to squeeze the maximum possible fibre from a given piece of land, government and industry are pinning their hopes on "intensive forestry," a practice entailing clear-cutting, tilling and slash-burning, followed by the planting of rows of single-species seedlings on the most fertile clear-cuts. With diligent weeding and thinning by hand or with herbicides, the approach produces plantations which yield fibre but which are structurally uniform and have lost much of the resilience of a biologically diverse forest to insects, disease, air pollution and climate change.

Just as the provinces have been reluctant to change their standard forestry practices to reflect new ecological knowledge, so they have been reluctant to acknowledge the current scientific uncertainty about how environmental change might affect the Boreal forest in the future. Although at least one federal study has shown that trees in the acid-sensitive Boreal forests of Quebec and Ontario are growing only one-third as quickly today as they did three or four decades ago, not one government is reducing its timber harvest today to accommodate the potential impact of acid rain or global warming on the forests. The growth of spruce and jack pine – prime commercial softwoods – dropped sharply after the onset of air pollution. A few decades from now, there could be far less timber on the stump than that projected by government statisticians.

Furthermore, new studies by Environment Canada estimate that even a slightly warmer planet, caused by a blanket of man-made gases trapping heat in the Earth's atmosphere, could shift the Boreal forest 100 to 900 kilometres to the north. Trees would grow more slowly in the south, and forest fires, insects and disease would attack the forest with greater intensity.

The irony of this scenario is that continued logging aggravates the global warming cycle by releasing stored carbon into the atmosphere. Indeed, one of the best ways to delay global warming is simply to leave the trees standing. Recent research shows that the forests in the Earth's northern hemisphere – one-quarter of which are in Canada – play a bigger role than previously believed in removing carbon dioxide, the chief "greenhouse gas," from the atmosphere. And although reforestation may seem an obvious answer to that problem, studies by the U.S. Forest Service conclude that replacing old forests with young plantations will do little to soak up excess greenhouse gases; even fast-growing seedlings will not absorb significant amounts of carbon dioxide for decades.

If Canada's Boreal forest is to be saved, say critics, government and industry alike must adopt new ways of thinking and acting – and soon. One of the first changes they suggest is in the current pattern of land tenure. To date, the Boreal forest designation as public land owned by the provinces has only encouraged pulp mills to monopolize a vital natural resource and replace it with uncertain monocultures. But the facts, says Lawrence Solomon, executive director of Environment Probe, show that much of the timber harvesting in Canada is simply not economical. Taxpayers often spend more money to furnish the industry with roads, cheap trees and fire and insect protection than they reap in stumpage fees, he says. Without such subsidies, pulp-and-paper companies would be less anxious to cut down remote forests because they

Where the Clear-Cut Is King

NOWHERE IS OUR EFFECT ON THE ENVIRONMENT MORE VISIble to the naked eye or disturbing to human emotions than in the toppled trees and ruined stumps of a clearcut. According to a national poll commissioned by the federal government in 1989, most Canadians disapprove of clear-cutting as a dominant method of logging. In addition, more than 80 percent report that they get upset when they see a large area of clear-cut forest. Not only do Canadians view such cutting as ugly, but they believe it to be harmful to the forest and its wildlife.

Although new research increasingly suggests that their sentiment is scientifically valid, clear-cutting – the practice of removing virtually all the trees from a site to obtain the most valuable fibre – remains king in Canada. Indeed, the practice constitutes 89 percent of all Canada's logging. In the Boreal region, big machines known as feller-bunchers plough into the forest, snip off trees like so many blades of grass, bundle them like kindling and deposit them at the roadside.

Economics, primarily the high cost of building logging roads, determines the size of clear-cuts. There are no laws and few provincial guidelines that limit their size. In some places, the bald patches left by the machines are so vast that they show up in satellite photographs. The largest continuous clear-cut in Canada is probably one southeast of Prince George, British Columbia, where nearly 500 square kilometres of rolling hills – an area five times the size of the City of Toronto – have been shaved clean.

The environmental and social costs of large-area clear-cutting include soil erosion, fatal landslides, floods and depleted fisheries. Such clear-cuts also destroy wildlife habitat and produce changes in the microclimate to which sensitive species cannot adapt. Cutting trees to the water's edge, for instance, was common practice in many provinces until biologists discovered that the resulting higher water temperatures had a detrimental effect on lake trout. Moose, bears and other large mammals need the cool temperatures in summer and the warm shelter in winter that large tracts of forest provide.

But smaller clear-cuts are not necessarily better. "There isn't a magic size for clear-cuts," says Daniel Welsh, a Canadian Wildlife Service biologist and expert on moose and songbirds. Limiting cuts to a fraction of a square kilometre, as the United States does, can produce a chequerboard of small, distinct forests in one area, each with its own age and species. Those animals best adapted to the artificial habitat would survive; but those with more specific habitat needs would be eliminated. And no future forest stand could be any bigger than the clear-cuts. When it comes to wildlife, "deciding that a 50-hectare clear-cut is going to solve all our problems," says Welsh, "is naïve."

–Christie McLaren

would not make money. "It's so unprofitable that the companies aren't willing to go in there without enough freebies from the government to make it worthwhile." With a little more imagination, Solomon argues, the industry could make the same return by investing elsewhere, governments could help create superior value-added jobs and tax revenues, and the forests could be given a new lease on life.

The answer, he says, is to shift the control of the land from the multinationals to plots owned by families. Private ownership not only fosters better stewardship but also changes the pace of cutting. In Sweden, where 75 percent of the forests are privately owned, small landholders are not cutting their trees fast enough to suit the government and the industry. Which, to Solomon, is as it probably should be. "If we had more control in the hands of ordinary people," he says, "their preference would be not to log. They would place a higher value on that land than the timber value. Clear-cutting, I think, would be quite unusual."

Solomon suggests that Canada start reallocating its Crown forests by settling all native land claims. Next, he says, it should set aside representative regions of wilderness and other ecologically important areas in parks and reserves. The remainder – which might not be a lot – would be distributed over 10 years according to a system determined by public hearings. The owners could do what they wanted with their forests, including timber production. Canada might lose its market share of forest products but would retain its watersheds and carbon sinks.

In response to Solomon's arguments, Louis Fortier of the Canadian Pulp and Paper Association is convinced the industry can sustain itself over the long term by practising intensive and extensive forestry. Anyone who suggests that governments should stop subsidizing regional forest development, he says, should apply the same arguments to the mining, fishing, tobacco and wheat industries. He also doubts that "Canadians would be prepared to give up their public ownership of the forest."

The challenge to the industry, however does not end with land tenure and subsidies. Another innovative approach to forest management is what its advocates are calling the New Forestry. It asks for nothing less than a revolution in forest management, an approach that would champion ecosystems instead of fibre factories. According to its proponents, which include senior officials in the U.S. Forest Service, it would do far less damage to the environment, while at the same time allowing timber harvesting to continue at sustainable levels. It would accommodate the traditional needs of native people, satisfy wilderness buffs and recreationists and cost less to maintain.

No more than two or three years old, New forestry has grown out of 20 years of studying the old-growth temperate rain-forests on the U.S. West Coast. A handful of scientists now agree that the key to a healthy, productive forest lies in the complex biological relationships that enable the old forest to regenerate new life continuously. "The legacy of the old forest must pass on to the new forest," explains David Perry, a professor of ecosystem studies at Oregon State University and a leading proponent of New Forestry. To do this, new logging methods must be devised to meet new goals. One is to protect the integrity of the soil, from which all life springs; another is to maintain the diversity of plant and animal life present in the original forest; and a third is to maintain genetic diversity within a single species of tree. Foresters of the future must try to preserve all visible and invisible components of the original forest, rather than eliminating them in favour of the commercial species. One of the most important organisms to protect, for example, is the mycorrhizal fungus, an almost invisible plant that lives underground but is crucial for helping trees to gather nutrients and water and as a defence against root pathogens. "We lose some of these critical organisms," says Perry, "and we lose the trees."

Logging, then, must be done in ways that mimic the natural patterns of disturbance in a particular type of forest. Instead of clearing a site of all dead trees and woody debris, for instance, snags and fallen logs could be left behind to rot, providing organic matter, nutrients and wildlife habitat and increasing the water-holding capacity of the soil. The new theory does not repudiate clear-cuts, but it would limit them to appropriate sites. A 200-square-kilometre forest managed under such a system might leave intact 30 square kilometres each of old, mature and young forest. Of the remaining 110 square kilometres, one-third on fertile flat ground might be one continuous, ragged-edged clear-cut with some standing green trees; another third, on hilly slopes, might be cut in narrow strips; the rest could be a mixture of selective harvesting, leaving small, open sunny patches, ragged patches or chequerboard squares. After harvesting, the area would be closed to further development for perhaps 100 years to allow natural processes to take over.

By keeping large forest regions in a more or less natural state, New Forestry would conserve wildlife, accommodate native people and sustain a large forest industry. But it would also mean an end to further industry expansion. "The expansionist mode is destroying the forest," says biologist Chris Maser, a leading proponent. "But if we redesign and scale down industry to match what the forest can produce, then – and only then – we can have an economically sustainable forest industry."

Canada's forest industry has not welcomed New Forestry ideals. "I think it's bunk," says Stephen Smith, vice president of timberlands for Weyerhaeuser's Saskatchewan Division. "It presumes that we haven't learned anything over all these years about how to grow trees and that strip cuts are better. I think we're smarter than that. We see no evidence to support it, so why should we do it?"

Ecologists reply that the future looks poor for the industry without far-reaching change. "The question is, do you bite the bullet now and try to manage forests in what some of us see as a more ecologically sound way, or do you bite it later?" asks Perry. "Society has to make that decision. We have to drop back to a sustainable timber yield, or nature is going to drop us back there."

Recently, a coalition of ordinary Canadians and northern local environmental groups proposed a moratorium on further development of Canada's dwindling forests, a proposal that may well be in order until Canadians have better information about the Boreal forest. Although the federal government recently created a full-fledged Department of Forestry – Forestry Canada – it is still labouring on a comprehensive "state of the forests" report to be tabled in Parliament next June. There is also a new recognition that existing forest research is inadequate for making decisions which will reach far into the future. In a major report in August, the U.S. National Research Council concluded that "forestry research must change radically" to encompass the new awareness of ecosystems in order to meet society's complex needs. Although the changes "will be expensive, difficult and, for many, painful," it says, "the consequences of not making them…would be more painful: a national and global society increasingly unable to preserve and manage forest resources for its own benefit and for the benefit of future generations."

As Russell Willier watches the clear-cuts move closer to his home each year in northern Alberta, he becomes more and more uneasy about the felling of his people's forest. If the destruction does not end by the next century, he predicts, Earth will rebel. Nature will try to heal itself, big winds will come, and a disaster like starvation could follow. The medicine man, who grew up in the forest, is among the first generation of Cree to worry about whether their children will enjoy the same entitlement. "The only thing I can do is train them that they can't depend on the forests no more," he says. "I have to; I have no choice." ■

This article was first published in the fall of 1990. Ms. McLaren is currently working on a book on forestry and the Canadian forest industry. Publication is anticipated in 1994.

Forestry Under the Microscope
Ontario's Search for a New Way in the Woods

Article by Jamie Swift

It resembles modern warfare. First, they send in the mechanized brigades. Then come the foot-soldiers. Aerial bombardment even plays a role. The entire operation depends on the latest technologies.

But the action isn't taking place in a dusty Persian Gulf desert. This campaign is being waged in Ontario's boreal forest, where the pine-scented stillness is usually disturbed only by the strident cry of the blue jay and the loon's gentle call.

The mechanized attack is led by diesel-powered machines, called feller-forwarders, that look like huge praying mantises. They rumble through dense stands of jack pine, their hydraulic cutting heads shearing off everything in sight and depositing them onto their backs. When fully loaded with tons of wood, the machines groan back to the roadside landing, where the trees are delimbed and dumped onto waiting trucks for the long voyage to the mill. Every year, the trip gets longer as the forest frontier recedes.

The infantry of industrial forestry consists of hundreds of little platoons of tree-planters, deployed in the wake of the logging machines. Every spring, this small army – mostly college students from the south along with a few local natives and whites – fans out across the north, packs stuffed with tiny seedlings to transform the massive clearcuts into productive forests.

Later, small planes fly over the new plantations, their specially fitted nozzles releasing a fog of chemical herbicide, that it is hoped, will kill off unwanted hardwood competition and allow the logged land to support a crop of spruce and pine whose long, strong fibres have been the basis for the success of Canada's single most important industry. One of the most popular herbicides among foresters is the old standby 2,4-D, a compound first developed when the U.S. military was looking for more effective chemical weapons during the Second World War.

Of course, such military images are hardly popular with the forest industry. Joe Bird, executive director of the Ontario Forest Industries Association, prefers the agriculture metaphor. He calls forestry a "cropping" exercise. Logging, which Mr. Bird describes as "an ugly word," is most always referred to as "harvesting" these days.

The industrial assumption is that yields can be boosted through judicious applications of the high-tech know-how. Intensive forest management is seen as both a source of future fibre and a way of reducing the growing political heat over wildlife and wilderness preservation.

"If in 50 years, we can double the yield per hectare, we cannot only increase the industrial exploitation, but we could make more forest land available for other purposes – deer yards or whatever," said Bud Bird, a former New Brunswick forest minister who also chaired a recent Commons committee on forestry.

Such optimism is based on an industrial concept of forests and forestry. Indeed, the intensive management of forests has increased almost as dramatically as the amount of forest land that has been logged over.

More and more wood is being extracted from Ontario's limited boreal forest than ever before. And once cut, forest authorities attempt to replace natural forests with plantations of commercially valuable species.

In the decade that ended in the mid-eighties, there was a 31.4 percent increase in the amount of Ontario forest being logged. Virtually all logging in Ontario is done by the clearcut method. Meanwhile, employment created by Ontario's forest products industry rose by only 4.5 percent and the number of jobs in logging actually declined to a record low in 1988.

A massive artificial regeneration program got under way in the early '80s, with three species (jack pine, black spruce and white spruce) accounting for 80 percent of all the trees planted. As more public money was pumped into forestry, the area thinned and "cleaned" also rose dramatically. And the amount of wilderness land sprayed with chemical herbicides also jumped. In 1976 one in every five hectares subject to treatment in Ontario was sprayed. Ten years later two of every three hectares treated got the herbicide once-over.

However, not everyone agrees that Canada's forests should be turned into huge fibre farms. Simplifying complex ecosystems by turning them into plantations is a risky business. Some argue that when forestry doffs its hat to modern agriculture, it is mimicking a system that has led to the impoverishment of soils, a fatal dependence on expensive chemicals and a breakdown of vital ecological relationships.

"This will end up leading to a biological holocaust," warns David Peerla, a forest campaigner for Greenpeace. "Mono-cultures simplify the biological diversity of the natural forest, thereby eliminating various plant, insect and animal habitats. We don't know how many of these species are being eradicated."

Disputes over the way we treat our forests are not new, but they have been growing more impassioned. A few short years ago, most Canadians had never heard of South Moresby or Temagami. Those names now evoke images all their own – blockaded roads, angry loggers and coffee table books featuring spectacular photos

Courtesy of Jamie Swift

of forests threatened by clearcutting.

Behind the headlines and the recent confrontations, the debate over forestry has changed. Until very recently, government forest landlords and their industrial tenants were struggling to convince the public that they were not mining the forests; that they were finally getting down to the tough, expensive job of setting in place a system of modern silviculture; that the forest was finally being treated like any other crop.

Now, however, critics of Canada's forest practices are no longer content to simply argue for the preservation of this or that spot of remaining wildland. Notions of biological diversity and ecosystem integrity have begun to appear. Something called "new forestry" has taken root and is spreading like ragweed.

One of the leading proponents of the new forestry is American zoologist Chris Maser.

"Nature designed a forest to be a self-sustaining, self-repairing entity. We're designing a regulated, economic plantation to require increasing external subsidies – herbicides, pesticides and, in many places, fertilizers," Maser told Ontario's Environmental Assessment (EA) Board. The board is conducting a detailed probe of the province's forestry practices.

Maser, who has become something of a guru to the new forestry movement, has a passionate, almost messianic approach to forestry. The former U.S. Bureau of Land Management scientist's testimony was the first that Ontario's EA board had heard that explicitly spelled out the new philosophy of forestry. He refers to nature with the same reverence and punctuation, that a cleric might bring to a discussion of the diety. "We have created the intellectual extinction of Nature's diversity through humanity's planning system, which invariably leads to biological extinction of species and their functions within the ecosystem."

Maser came to Ontario earlier this year to testify on behalf of Forests For Tomorrow, a coalition of environmental groups. The coalition is challenging Ontario's Ministry of Natural Resources and its industrial clients, whose message to the EA board is quite simple: *All's well in the woods.*

The quasi-judicial hearings, which got under way three years ago and proceeded quietly in an old railway hotel in Thunder Bay before shifting to Toronto, will be crucial in determining the direction of forestry in Canada. Michelle Swenarchuk, the lawyer representing the coalition at the hearings, says that their importance lies in the fact that, unlike a royal commission, the two-person EA board will not make recommendations. They will make a decision, in effect rewriting Ontario's forestry rule book.

But the hearings have heard sharply different views on how the forest should be treated.

"I would say that there's been a complete failure on the part of industry and government to acknowledge the need for any change of practices," says Swenarchuk, who has learned a lot about the importance of forest ecology since she started work on the case in 1986.

The former labor lawyer sits in her small office in a converted factory and warehouse not far from Toronto's declining garment district. The impressive beams, at least 16 inches square, are cut from white pine, a species that was the backbone of the eastern forest industry in the days when industrial promoters believed the forest was inexhaustible. Today, the forest industry seems equally confident that its-clear-cut-and-plant strategy is the way to sustain its mills.

"In the two years it took for the Ministry of Natural Resources and the industry to present their cases, we heard scarcely a word about concerns over full tree harvest and nutrient depletion," recalls Swenarchuk, her voice rising. "There was no reflection of American thinking about new forestry or the views of German scientists worried about plantation management, artificial regeneration techniques and our ability to even grow a future forest that way. In fact, we were treated with *utter contempt* for raising these concerns until the volume of the evidence we presented and the strength of the literature indicated that this is not an aberrant approach."

The question of full-tree logging and nutrient depletion provides a glimpse at the chasm separating advocates of industrial forestry and those promoting the new forestry.

Full-tree logging is a highly mechanized technique that involves extracting the entire tree – branches, top and all – from the site and hauling it to a roadside landing, where it is limbed and topped. This "waste" is left to rot in large piles. The technique differs from older, tree-length logging systems in which the tree's branches and top are removed where it is felled.

Until recently, tree-length logging systems that leave the nutrient-rich foliage on site accounted for 70 percent of the cut in Ontario. But Ministry of Natural Resources officials told the EA board that tree-length logging has declined to about 30 percent. The dominant system is now full-tree logging because of corporate concerns over the cost of wood. "The way logging methods have developed in Ontario...has been largely influenced by economic considerations," ministry officials admitted.

The tree tops, small branches and foliage left after logging, referred to as slash by foresters, plays an important role, serving as an environment where decomposers immobilize large quantities of nitrogen. This reduces losses of nitrogen and serves as a nutrient sink to feed new growth. Industrial forestry's concern with removing wood cheaply has led to the dominance of full-tree logging systems and the removal of valuable nutrients from the soil.

Michelle Swenarchuk is alarmed by this approach. "It's obvious even if you don't know any science. When you take the whole tree off the site, you're removing not only the wood, but also a great deal of material that would decay to help form a nutrient base for the next generation."

The contrast between commercially driven forestry and the new forestry is striking. The dominant regime, based on artificial regeneration and intensive silviculture, concentrates its research on clonal forestry to produce new versions of the tree species most in demand by pulp mills. Joe Bird, of the Ontario Forest Industries Association, is optimistic about all the new seed orchards that will soon produce genetically manipulated trees.

"Forest managers in industry and government see it as a great opportunity. The tree improvement thing is quite a success story."

The official voices of industrial forestry focus on numbers. They count the number of hectares planted, the amount sprayed with herbicides and the amount being "managed" by commercial foresters. This involves a massive, unprecedented alteration of the boreal forest ecosystem, a publicly funded experiment of enormous proportions.

Adam Zimmerman, chairman of Noranda Forest and the most outspoken defender of current practices, bemoans the industry's poor environmental reputation.

"Clearcutting isn't pretty," says Zimmerman. "But modern clearcuts are of modest size and are usually scattered...The fact is that Canada's forest is being cared for better than ever before."

For Adam Zimmerman, a well-cared for forest is a "managed forest."

"But what does the managed forest mean? It means that when a forest is harvested, seedlings have to be planted and cared for on lands that have been prepared for plantation."

The head of Canada's largest forestry firm points to the 900 million seedlings planted annually, more than triple the number planted in 1980.

"Industry will adamantly point to the billions of trees they've planted," says Swenarchuk. "And I'm sure they have.

The issue is that we should not be relying on large-area clearcutting and planting at all. The emphasis should shift very substantially away from clearcutting and planting to cutting small areas to enhance natural regeneration."

Such thinking is very much in line with the tenants of the new forestry. And while the forest giants exhibit a can-do mentality, confidently predicting success in their new plantations, the new forestry has more modest expectations. Not the least of these is the need to reach an understanding of how forest ecosystems work before attempting to replace them with fibre farms.

This is Chris Maser's perspective. It is radical not only because it is such a significant departure from conventional forestry, but it also literally entails getting at the root of the matter.

This is why Maser has spent so long studying squirrel turds.

Maser refers affectionately to such droppings as "calling cards." When small mammals eat sporocarps (fungus fruiting bodies), they nourish themselves and also help to build the links in the chain of ecological relationships that allow the forest to thrive. Each fecal pellet emitted by such foragers is what Maser calls a "pill of symbiosis," containing nitrogen-fixing bacteria, yeast and the spores of hypogeous mycorrhizal fungi, tiny hair-like tendrils that grow underground.

The mycorrhizal fungi are important to forests because they form sheaths around the fine roots of trees in a symbiotic relationship, taking energy from the trees but helping them process nutrients from the soil. Some mycorrhizae also prevent pathogens from contacting the root system of the trees.

"If all this sounds incredibly complicated, that's because it is. And it is only one tiny glimpse of a forest's total complexity," Maser told Ontario's EA board.

To what extent can massive clearcuts such as the 200,000-hectare one that Forests For Tommorow discovered outside Kapuskasing ever support a viable forest? How do small mammals react to habitat changes in the 20,000-hectare clearcuts that are not uncommon in northern Ontario? Which species will become extinct and what is their role in the natural functioning of the forest? What will be the effects of the plantation strategy on the biodiversity of the boreal region? What is the extent of that diversity?

Since no one really understands the dynamics of boreal ecosystems, and since massive plantations have never been tried before in northern Ontario, it is difficult to confidently claim that these forests are being managed in a sustainable fashion.

Ontario's four-year environmental assessment of forestry will provide some clues about what sustainable forestry will mean in Canada. When it finishes its deliberations, the EA board will have conducted the most detailed and wide-ranging study of forestry ever undertaken in Canada. And all the participants – industry, government, native groups, environmentalists, hunters and anglers – know the board has the power to issue a finding that could alter the way forestry is practised.

Last September, an unexpected factor was added to a political equation nearly as complex as the forest itself. Ontario's New Democrats surprised everyone, including themselves, by forming a majority government.

This was a party that had for years been highly critical of the Ontario's forestry practices. But the NDP position was not taken too seriously by those in control of the forests. On Michelle Swenarchuk's cork bulletin board hangs a copy of the cover of *The Last Stand,* an NDP study of Ontario forestry, published in 1983. She got the report from the Ministry of Natural Resources library. The acquisition date, rubber-stamped beside a line drawing of a conifer, is September 7, 1990. The NDP was elected on September 6.

A key element of NDP forestry policy has always been a shift in control of the forest, with more power transferred to those closest to the resource (the NDP has a strong northern caucus) and away from southern-based bureaucracies. "The forest looks far different up close than it does from the Whitney Block or the corner of King and Bay," the party was fond of repeating.

Sitting in his spacious corner office on the top floor of the Whitney Block is one of the people who put together *The Last Stand.* But Wildman is Ontario's new minister of natural resources.

Asked if Ontario's dwindling reserves of natural forest can sustain ever increasing cut levels, Wildman replied bluntly, "No, it can't. Not if we continue going the way we're going."

The MLA for Algona, a huge riding north of Sault Ste. Marie, is conscious of the debates raging over new forestry and a growing realization among many European foresters that their centuries-old faith in plantation forestry has yielded biological deserts, where both diversity and productivity are declining.

"In Scandinavia and in places like the Black Forest of Germany, where they've been far more successful than we have in the past in terms of farming the forest, they have found that after a long period of time, single crop monocultures do not even provide quality fibre. They're now moving to look at questions related to biodiversity."

Under the NDP, Ontario has taken a few tentative steps to change its approach to forestry. Aerial spraying of herbicides has been cut by 20 percent. The ministry of natural resources has published a formal document outlining the principles underpinning its version of that slippery concept, "sustainable development." They include:

"All life is connected, from the fungi in the soil to the birds in the sky. Human activity that affects one part of the natural world should never be considered in isolation from its effects on others...Our understanding of the way the natural world works – and how our actions affect it – is often incomplete. This means we must exercise caution and a special concern for natural values in the face of such uncertainty..." it states.

But how will this translate into real changes in the way the forest is treated? Crucial issues such as clearcutting, full-tree logging and the province's plantation strategy in the boreal forest are still unresolved. The NDP is adopting a wait-and-see attitude. They've commissioned a state-of-the-forest audit, an old-growth study and plan to look at pilot projects in community-based forestry.

"There is a tremendous lack of information and data," Wildman says. "We don't even have much information about what's out there in the boreal forest."

In spite of the fact that little has changed, the new regime has high-sounding hopes. The minister in charge of Canada's third-largest forest, with 80 million hectares of forest land, believes he can engineer a turnaround. Indeed, Ontario's formula approach to forestry, characterized by the clear-cut-and-plant approach, was subject to a lengthy 1986 investigation by Gordon Baskerville of the University of New Brunswick, one of the most respected figures in Canada's forest establishment. Baskerville dubbed it "Betty Crocker Forestry."

"We now have what we call timber management plans," says Wildman, rolling his eyes. "I hope we'll be moving to *forest* management plans. All of the timber management units are completely arbitrary; they should be on a watershed basis. We're talking about integrated management in which you're going to have wildlife biologists have as much say as the forester in the development of the management plan."

This sort of talk has Ontario's big pulp companies edgy. Joe Bird worries about "the politicization of silviculture."

There are powerful interests with large investments of financial and intellectual capital in the present way of treating the forests. The forestry giants can be expected to resist any changes that will raise the cost of timber or their access to supplies. Within Wildman's own ministry there are many foresters who regard the new forestry with profound suspicion.

Industry balance sheets have been ravaged by the downturn in business. The pulp industry has historically failed to keep up with its innovative competitors in Scandinavia and the United States. Ontario's traditional newsprint market south of the border is threatened by new regulations demanding a higher content of recycled fibre than northern mills can easily supply. And demand for newsprint is stagnating in comparison to demand for speciality papers.

These factors, coupled with the threatened closure by Spruce Falls Power and Paper – the only major employer in the clay belt town of Kapuskasing – and shut-downs by Abitibi-Price in Thunder Bay, mean that the private sector is well positioned to resist moves it regards as interference with the way it treats the public forest. The job blackmail card is difficult to trump. And the forest industry will resist moves to reduce clearcutting because that could raise the cost of wood.

Ontario's reassessment of its forestry practice and the new government's stated intention to do things differently is being played out against this background.

Activist lawyer, Michelle Swenarchuk regards the NDP's rethinking of boreal forest practices as "very positive." But, she is still wary. "The question is: what are they going to mean in practice? How much will be implemented? How will they take on the key issue of harvest method and regeneration policy?"

Only time will tell if the future of Ontario forestry will simply mirror its unhappy past or whether some real changes will emerge. But it's clear that in Ontario and in other provinces it's not too late to learn from past mistakes by adopting ecologically sustainable methods of treating the forest.

"Canada should skip the intensive forest management stage that has been tried and found wanting in Europe," says David Peerla of Greenpeace. "We should move to preserve as many remnant biotopes (small, distinctive ecological communities) as possible, reduce production and consumption of one-use paper products and modify our forest practices so that they are as gentle on the environment as possible. There are huge uncertainties about the consequences of our current practices." ☐

BY DOUGLAS COWELL

The Winds of Change Blow Through Community Forests

Across the country the idea of community forests is gaining respect.

Question: What does the recent federal referendum and community silviculture have in common? Hint: Think "mood."

Amongst a certain crowd within the forest sector in this country "community forestry" has become a hot item during the last three or four years. Silviculture spoke with several contacts coast to coast. Two things quickly emerged as common factors.

First, although everyone seemed quite aware of the term, almost all asked us, "What's your definition of community forestry?" There are actions, plans, pilot projects, surveys, policies and, in a few cases, actual forests across the country that can be linked, jammed or gerrymandered into the classification of "community forestry." What there isn't, is a common definition of the term.

Let's pick a ready made one. Dr. Julian Dunster, a professional forester and environmental consultant has studied community forestry around the world.

"The true community forest is a community tenure, based in legislation. Such a tenure grants the community rights to manage the whole forest in an integrated manner, meeting goals that are defined by the community within a broader regional and provincial framework. It is not yet a community forest if the ultimate control over the land remains subject to provincial veto."

By that definition there is very little community forestry practiced in Canada.

In some ways that's not too surprising, considering our history and how recent the upsurge in interest in the subject is. As Canada developed, private and corporate interests plus higher levels of government staked their claim on forests. Ownership and tenure migrated to these groups and most people neither noticed nor cared that communities were generally without control or influence over their forests.

Across Canada, and indeed, around the world according to Dunster, that's changing fast. Peter Sanders is a lecturer and resident silviculturist at the University of British Columbia Research Forest in Maple Ridge and is president of the B.C. Federation of Woodlot Associations. He recently presented a paper at a public workshop on community forestry in Quesnel, B.C. It was sponsored by that city's economic development branch. Sanders is organizing a meeting on the same subject in March of 1993 for professional community planners and consultants. He's very aware of what's going on in British Columbia.

"There are people right across this province with a fair amount of enthusiasm for the subject," he reports. "Communities want a greater say in their viewscapes, resource use allocation and who does what."

In fact, it's Sanders that asked and then answered our opening question about the referendum. "Most Canadians, I think, saw a lot of good things in the Charlottetown accord. But they voted "No" because people want more local and community decision making."

A big statement but he backs it up with the finding of the B.C. Forest Resource Commission, positions taken by Forestry Canada and the hearings of the federal Standing Committee on Forestry. All report people are demanding more direct say in the management of their forests.

Let's take a look at the two examples of community forests that do meet Dunster's demanding requirements. Two small communities in British Columbia, North Cowichan on Vancouver Island and Mission in the Lower Mainland, have working, municipally controlled forests. Both acquired the core of their forest land in the 1930's through non payment of taxes.

Darrell Frank is the municipal forester for North Cowichan and is obviously delighted with his 5,000 hectare domain. "Because of the foresight of people back decades ago we now have a working forest that we manage both for timber and a wide range of other considerations.

"The population of our area is growing tremendously. Our municipality is about 20,000 hectares in all but we own 5,000 as forest. People can use if for wildlife viewing, hunting, water supply, as an educational tool, and it creates jobs too."

North Cowichan began managing its forest in the 60s simply by letting small contractors in to cut. Concern in the late 70s by then-mayor, Graham Bruce, led to the creation of a much more sophisticated multiple use plan that, assisted by a government Employment Bridging Program, resulted in a large backlog of thinning, pruning, spacing and fertilization being completed.

In 1982 the municipality hired its first professional forester. Today the forest is managed for integrated resource use. Small clearcuts are used and cutting, tending and planting is all done on contract. Their profits from log sales have been plowed back into roads and other improvements and into a reserve fund, all without cost to the community.

The City of Mission's forest is double North Cowichan's at about 10,000 hectares. Most of Mission's land is leased from the provincial government and is managed under a Tree Farm License, the same long term Management tenure granted to major corporations. But as Kim Allen, a professional forester and the Director of Forest Management for Mission explains, their goals and methods are much the same as Cowichan's.

"We cut about 40,000 cu. metres a year and do all the usual planting, brushing, weeding, spacing and protection. We use widely scattered, mostly small clearcuts from less than one hectare up to about 40 hectares but we're also trying some horse logging and thinking about commercial thinning.

"We want to manage according to community values and always consider landscape aesthetics, environmental concerns, employment, recreation, and revenue."

Since they began managing the forest in 1958 they have always made money for the city, except during the recession of the early 60s. In the late 80s they not only met their goal of providing $800,000 a year to city coffers but also built a $2 million reserve fund and contributed an additional $500,000 to help build a firehall and buy a fire truck.

"It's also a particularly good education tool," Allen explains, "especially for local politicians who get a good grassroots education in forestry."

That theme of education is common across the country. In Newfoundland, for instance, a province that seems to have missed out on most of the community forest fervour, Forestry Canada has just established the Thomas Howe Demonstration Forest on Department of Transport lands adjacent to the Gander airport.

Senior Development Officer on "the rock", Bill Alexander explains that it was set up mainly for educating school children but he makes a comment that indicates the interest is wider than that: "People are taking a go slow approach right now. They know they're supposed to do something but they're not sure what it is." The demonstration forest seems a good start and as he says, "When you get people involved they understand much better what you're trying to do."

In Nova Scotia, too, where 75% of the forests are privately owned, 50% by thousands of farmers and smallholders, the song of community forestry is also being sung. The context is much different than in Western Canada, for instance, and it's not very likely that Dunster's definition will be met in many situations. None the less, there are programs to help land owners manage their woodlots cooperatively and market their wood more successfully. Education for both landowners and members of the local community is key according to Forestry Canada's Ian Millar.

British Columbia is perhaps the hotbed of community forestry, with various communities, organizations and even citizens groups on the Gulf Islands all wanting their own forests. Interesting certainly, but it's not necessarily surprising. 'Lotus Land' is a magnet to all of Canada's innovators, misfits and malcontents. As one New Brunswicker said, "We consider that all new ideas in forestry begin in B.C. and move east."

Surprisingly, Ontario may turn out to be the big innovator in community forestry. Stephen Harvey, Senior Policy Adviser in the Forest Policy Branch of the Ministry of Natural Resources, explains that there's a new direction in sustainable forestry being undertaken by the ministry.

"We are looking for ways to increase community involvement."

He underlines that community means the broader community and not just municipal and local governments. He also emphasizes that involvement means active, effective involvement and not just emasculated "input" too often requested by bureaucrats and politicians.

To the outside observer the winds of change in Ontario seem to be blowing at hurricane force for a province that has just come through a three year forest management Class Environmental Assessment program, whose main thrust seemed to be to prevent involvement by environmental organizations, citizens and other trouble makers.

The general impression is that the ministry is proceeding exceedingly cautiously – tippy-toeing on pins and needles. But they are proceeding.

To begin with, they commissioned a review of all of Ontario's experience in community forestry. Then they created pilot projects with four communities: Geraldton, the 6/70 Area Economic Diversification Committee of Kapuskasing, the Township of James (Elk Lake) and the Wikwemikong First Nation whose reserve is on Manitoulin Island.

Each of these has been funded to develop a community plan (basically each community gets to choose how much and which forest it's interested in), implement their plans and then review the experience. The ministry is trying to build up a base of experience on which to found a policy for community forestry in over 100 communities.

The scope of the "new direction" being taken is staggering. There are however, a few restraints. First, in every case the community includes the provincial government and whatever forest industry is involved locally. Second, there is very little productive forest land in the province not already committed under long term lease. Interestingly, Harvey doesn't expect a conflict between community plans and big industry. "Generally companies have been supportive."

Why is this happening in Ontario? "There's a demand by the public for more input and control," says Harvey. "I think that's the case in any government activity and forest management may be a little slower than others."

Pushed a little he suggests carefully, "There is definitely a different mood in the province; one in which the current government is setting the direction for forestry."

"That mood, in varying intensities, extends across the country and if it prevails we can expect some serious changes in community forestry." ∎

SMALL IS BEAUTIFUL

MERV WILKINSON HAS BEEN MANAGING HIS LITTLE WOODLOT SUCCESSFULLY FOR 50 YEARS

IN 1936 MERV WILKINSON BOUGHT 55 hectares of land at Yellow Point, a peninsula just south of Nanaimo, BC, on Vancouver Island. Hoping to turn it into a farm, he went off to the University of British Columbia to study agriculture. It was there, through the influence of one of his professors (who happened to be a forester trained in Sweden), that Wilkinson decided to learn forestry instead. After graduation, he named his property Wildwood and began to manage the land as a woodlot.

Wildwood is a perfect example of the dry coastal Douglas fir forest. Virtually all species of trees indigenous to eastern Vancouver Island are thriving there; there are even some Douglas firs that Wilkinson figures are over 1800 years old. The climate is moderate, the terrain is gentle (making roads easy to build), and it is close to lumber markets — a forester's dream.

Wilkinson's dream was to manage a coastal forest that would produce wood products without losing all the characteristics he had come to love and respect. His vision must have been a clear one because today he has a healthy, thriving woodlot that has provided him with about a third of his income over the last 50 years. And despite cutting approximately 1.4 million board feet during that time, Wilkinson figures that — prior to making his ninth cut in 1990 — he had the same amount of timber standing as he did when he began in 1938.

Wildwood is living proof that you can have your forest and mill it too — at least on the small scale. Its name is especially appropriate; the property resembles a wilderness forest, even though human management is always giving nature a gentle push to produce high quality wood products. Wilkinson estimates that he has increased the growth rate of his trees by an average of 12 percent.

The forester attributes his success to "common sense" and "working with nature." One technique he uses employs the basic principles of genetics. Trees with desirable characteristics — tall, straight trunks, abundant foliage, and good cone production — are left throughout Wildwood to provide seed for future generations. This is the best way to ensure that there will be a steady supply of healthy young trees that are genetically adapted to the site they germinate in. While other practitioners of "selective logging" often send these prime trees to the mill and tally up the profit right away, Wilkinson sees a larger (albeit long-term) profit in leaving the parent trees to do what they do best — maintain the quality of his forest in perpetuity.

This approach to working with nature takes patience, and Wilkinson has spent a lot of time observing the trees on his woodlot. For example, he has recognized that in certain parts of his forest the cedars and firs grow in cycles. "When I first came here there were cedars. I took them out and they were replaced by firs. Now I'm cutting the firs and the cedars are coming again. This is nature's way of ensuring against disease and soil depletion," he says. He points out that in this country foresters get little chance to have a long-term relationship with the area they are working in because they are moved around so much. "In Europe," he adds, "they have more continuity — some of them stay with the same forest for 40 years or more."

Wilkinson also uses animals as indicators of conditions in the forest. While a lot of foresters will take out a tree as soon as it's dead because they feel it is a hazard, he relies on woodpeckers to tell him when it's time to take out an old snag. "You can see them move their nests down each year as the snag rots. When they abandon the tree altogether, it's time to take it down — it's not safe anymore." In the meantime, he says old snags can be very beneficial; they provide homes for insects, birds, spiders, and other species that maintain the balance of life in the forest and pre-

Courtesy of Marion Lightly

vent outbreaks of destructive insects.

The BC woodlot owner doesn't do any clearcutting either, nor does he burn slash or use chemicals to remove undergrowth. The canopy is left intact in order to keep the soil shaded and moist; but he opens it up through thinning, to let in enough light to encourage the growth of young trees. Limbs and debris are scattered about on the ground to replenish the soil, and he practices some brush control to cut down on competition with the growing seedlings.

There is a delicate balance here, he says, because in some ways the brush is beneficial. It protects the seedlings, harbours birds and wildlife, and can replenish nutrients to the soil. But it can also be a hindrance when it competes with the very young seedlings for light, water, and nutrients. He has learned through the years that sheep, given supplemental feed and limited access, are very effective brush control agents. Unlike deer, they don't seem to like eating the young conifer tips. But Wilkinson only lets them graze while the seedlings are small — once the trees are free growing, he removes the animals and leaves the brush alone.

Of course, the forester and other practitioners of selective logging are not without critics. Clearcutting advocates say such methods don't open up the forest enough to allow healthy regeneration of young trees. In response, Wilkinson points out that trees grown in the open have trunks with a lot of thick lower branches, widely spaced growth rings, and a conical shape — all of which produce poor quality lumber, if they're used for lumber at all. Most end up as pulp. In contrast, saplings grown among taller trees have to stretch for the light in their early years and end up with tall, straight trunks and fewer and thinner lower branches. This leads to tight-grained, knot-free wood with a much higher market value.

Two other arguments against Wilkinson's methods relate to sustainability. One is that selective logging which uses skidders to haul out the logs (as his does) will result in soil compaction and damage to roots after five or six passes. This, of course, only applies to the trees whose roots are under skid trails which are used over and over. Although evidence of such damage is not immediately visable, it could show up as die-back at some point in the future.

The other major concern is root rot. This is a condition caused by fungi inside the root tissues — the two common species on eastern Vancouver Island are *Phellinus weirii* and *Armillaria mellea*. While it is common for the roots of apparently healthy trees to be infected, the fungi are normally in a suppressed condition. However, when a tree is cut down, the fungi spread to neighbouring trees, wherever there is root contact.

While "anti-selective" foresters warn of potential outbreaks of root rot in Wilkinson's forest, he is probably safe. Each species of fungus has a preferred tree host, so where there is a mixture of tree species, as is the case at Wildwood, the spread of this condition is kept in check. Also, even in clearcut operations, the roots are left in the ground where the fungi responsible for this condition may remain infectious for decades.

The question now being debated in BC is whether selective management can be successful on a large-scale commercial basis. David Handley, a forester for MacMillan Bloedel, has prepared papers on why selective logging would not work in BC. While the arguments sound convincing, it is easy to see that the economic issue overshadows any biological or ecological considerations. Selective logging is more costly in the short term, especially road-building, and profits are a lot further down the path.

Even though Wilkinson has been managing Wildwood for 50 years with very promising results, it's still too early to tell if his initial success will continue indefinitely. Perhaps in another 100 to 200 years we'll know the answer. But in the meantime, he is providing a practical model of forestry that is a real inspiration to those who despair at our vanishing forests.

MARION LIGHTLY

Discussion for Forestry

1. The movement toward "New Forestry" is based on the obvious need to exploit forests sustainably – that is, to regenerate what you cut in its full biodiversity and timber potential. Discuss the strategies being offered by proponents of New Forestry, why each strategy is good for the forest, and why forestry companies might resist such measures.

2. The use of chlorine in industrial processes is a major environmental concern because of the toxicity of its byproducts. Pulp mills across Canada that use chlorine to bleach pulp are flushing dioxins and furans – highly toxic byproducts of chlorine bleaching – into rivers, lakes, and oceans. There are studies that show these toxic wastes harm wildlife and are a health hazard to humans even in minute quantities. Because they don't break down readily, dioxins and furans also tend to accumulate in plants and animals, thus growing in hazardousness over time. Only a few years ago, pulp and paper companies claimed bleaching without chlorine was not technically feasible. But today in Europe and to a lesser degree in Canada, progressive pulp-and-paper companies are using chlorine-free bleaching technologies. Should these technologies be mandatory under law in Canada? What if pulp companies have had financial difficulties and say they can't afford the upgrade? What if they say jobs will be lost if they are forced to comply with the law? How can pressing environmental concerns like organochlorine contamination be balanced with economic imperatives?

3. Discuss the possible benefits and drawbacks of public ownership versus private ownership of resources like forests.

4. By leasing public lands to large foreign companies that export pulp and logs to their own countries, Canada is only deriving revenue from "raw" resources. This is far less than the revenue earned by those same companies when they convert the logs into lumber and pulp into paper. Is this right? What does this say about Canada's economy and its economic future? And should we be concerned that foreign companies, determined to earn maximum returns on the forests they lease, may not be good environmental stewards? If forest stewardship is an added expense, wouldn't they do all they legally can to avoid it?

5. Consider the term "forest management." What does it mean to: 1) forest companies, 2) governments, 3) logging communities, and 4) conservationists?

6. Again, we are confronted with the question of scale. As with the East Coast cod fishery, the forests of Canada are harvested by machines designed to take as many trees with a little human labour as possible. They are efficient from an economic standpoint, but they leave vast clearcuts that destroy forest ecosystems. This is centralized forestry as practiced by large forest companies. At the other extreme you have Merv Wilkinson, who has earned an income from sustainably managing his 55-hectare woodlot for the past 50 years, and today there is more timber on the lot than when he started. Most people would believe that small-scale forestry like Merv practices or like horse-loggers are practicing is impractical, but there are plenty of small-scale logging operations in Sweden and Finland that supply logs to large mills for processing. Discuss the appropriateness or inappropriateness of small-scale, decentralize resource extraction. How might it help the environment? How might it help Canada's unemployment situation? Why do you think it is not practiced more in Canada? Could we make a transition to many small owner/operators from the large companies that harvest logs today? Would forest companies resist a move toward small-scale logging?

7. Why do you think the community owned forests have not been overexploited? The benefits of community owned forests are measured in more than money. Discuss the total benefits. All these benefits added together are known as "true-cost resource accounting." Consider the true costs of clearcutting huge expanses of public forests. If money for trees is the "revenue" one side of the balance sheet, what are the "losses" on the other side?

Introduction

The James Bay Hydroelectric Project

This issue examines the impact of the first phase of the James Bay hydroelectric project on the native people who live in the region; it shows the position of the Canadian government and Hydro Québec on the proposed second phase of the project; and why many people are opposed to the building of more dams in northern Québec.

Before the 1970s, James Bay was known by most Canadians as the southward extension of Hudson's Bay. Today, mention James Bay and many people think of a controversial "megaproject" in northern Quebec.

In 1971, the Quebec government announced the impending construction of La Grande, the first phase of its plan to transform some of the mightiest rivers on the Quebec side of James Bay into a massive hydroelectric power complex. By the time construction was complete a decade later, Hydro Québec, the publicly owned electric utility, had diverted seven rivers into the La Grande River, built 215 dams along its course, and channeled the water through dozens of penstocks, turbines, and draft tubes to yield 10,000 megawatts of electricity. (That's just under half the peak demand for electricity required by the entire province of Ontario.) The big dam at La Grande 2, the pivotal site of the project, is as tall as a 53-storey building and its powerhouse is the world's largest with 16 one-thousand-tonne turbines.

Hydro Québec and the Quebec provincial government have always insisted that the James Bay project is essential to the province's economic future; that the power produced there will not only supply Quebec's own industrial growth, but will be an important export commodity when sold to American states and other provinces within Canada, earning Quebec a guaranteed income for many years to come. They also say hydropower from James Bay is environmentally preferable to any alternative method of generating such high volumes of electricity, such as nuclear fission or the burning of fossil fuels.

But environmental analysts and activists disagree. They are opposed to the second phase of the James Bay project, known as The Great Whale in reference to the main river it would affect. They say the La Grande phase altered the wilderness terrain too severely, contaminated the river water with toxic mercury, and worst of all, effectively ended the traditional lifestyle of the indigenous peoples. Furthermore, they have demonstrated that the power produced by Great Whale is simply not needed.

Native peoples living in the James Bay region – specifically the Cree and Inuit of northern Québec – were not informed in advance about the first phase of the project. The provincial government made a public announcement and proceeded almost immediately to build roads that would drastically change native lives by penetrating for the first time pristine areas formerly accessible only by boat or floatplane. Through the courts the Cree and Inuit tried to stop what they considered an invasion of their territory but gave up after four years, believing opposition to be futile. In 1975, native representatives signed the James Bay and North Quebec Agreement with the Government of Quebec, the Government of Canada, and Hydro Québec. In exchange for certain land claims relating to the proposed hydro projects, the Cree and Inuit earned special hunting, fishing, and trapping rights; a greater measure of self-government; and $225 million to be paid over 25 years.

The social and environmental problems that resulted from that agreement have caused the natives of James Bay to fight the second phase of the project with far greater resolve. Well organized and well advised by public relations experts, lawyers, economists, and environmentalists both in Canada and internationally, they have forced an environmental impact study by the federal government that is said to be the most rigorous ever conducted for a major development project in Canada. (The second phase of the project will convert over 80 per cent of the Great Whale River into reservoir; 4 dams and 29 dykes will be constructed; two other rivers will be diverted to increase flow; and the work will require 2 roads, 3 village/workcamps, and 3 airports.)

Megaprojects like La Grande and Great Whale inevitably disrupt the communities that must live with them, and Canadian aboriginals are especially hard hit because of their lifestyle. As Jamie Linton writes in Nature Canada (Reading #1), "Because the Cree have been so closely wedded to the land, the James Bay project affects them in a way that is hard to imagine. Northern Quebec's rivers are the foundation of the traditional Cree economy. By altering them and flooding land to create huge reservoirs totaling half the surface area of Lake Ontario, the Quebec government has turned the lives of the Cree upside down."

The Cree town of Chisasibi at the mouth of the La Grande River shows how devastating the changes have been. Before 1981 it was called Fort George and was actually located on an island at the river's mouth. The inhabitants were relocated on shore because the increased volume of the La Grande washed their former village into the bay once full flow was achieved. The construction of a highway ushered vehicles and outsiders into a settlement that had never faced them before. The people now suffer alarming rates of alcohol abuse, divorce, teenage pregnancy, and suicide. Economically, Chisasibi has bleak prospects. Only 41% of the population of 3,000 is able to make a living from hunting and trapping. There is no economic base to support the rest. While thousands of jobs were created by the construction of the hydro project, only five Chisasibi residents now work for Hydro Québec. "[T]he people are being forced to undergo prodigious change at an impossible pace," writes Linton. "It is as if the TV-watching, car-driving, supermarket-shopping inhabitants of our cities were compelled to quit their jobs, demolish their apartments and houses, and go into the bush to fend for themselves."

While hydroelectric power might be environmentally preferable to burning fossil fuels, it is not without its detrimental effects. In the early 1980s two-thirds of the residents of Chisasibi were shown to have unacceptably high levels of mercury in their bodies. The mercury, it turns out, was a byproduct of the flooding. Scientists have discovered that a bacteria released by the water converts naturally occurring mercury into poisonous methyl mercury. Concentrated in the food chain, this toxin accumulates in humans who eat fish from contaminated water. So today, a people who earned income from fishing and have eaten fish as a staple of their diet for centuries have been warned by the provincial government to reduce their consumption. Women of childbearing age have been told to stop eating fish altogether.

The Cree and Inuit of James Bay have earned the sympathies of the international environmentalist community, including the influential National Resources Defense Council (NRDC), a New York-based lobby-and-advocacy group staffed largely by lawyers and economic analysts. Reading #2 is comprised of letters published in *Amicus Journal*, the NRDC's magazine. The first two letters are written by Canada's ambassador to the United States and the chairman of Hydro Québec. Ambassador Burney defends Canada's signing of the James Bay and North Quebec Agreement and emphasizes that the environmental review of the second phase of the project will take into account the impact of further construction on native people and the environment. Richard Drouin of Hydro Québec says the extra power from Great Whale is required by the province; it is the safest form of power; and that Hydro Québec's role in mitigating the effects of mercury contamination have been unfairly downplayed. The third letter, from NRDC staff members, point outs that the environmental review of the project is not, as the ambassador and chairman seem to imply, a goodwill gesture on the part of the governments of Canada and Québec, but rather the result of numerous lawsuits that forced the issue; that the mercury exposure of people in the region is dangerously high and mercury levels will not likely abate for over 90 years; and that power from the Great Whale phase is in fact not needed – that if Hydro Québec were fully committed to energy conservation and "demand management," it could save more power than it intends to produce, at a much lower cost.

As Pegi Dover and Philip Jessup express (Reading #3), the most progressive electric utilities in the world are now relying on conservation and demand management for their future power requirements. Indeed, energy conservation was the key factor in the decision by the New York Power Authority to cancel a major contract with Hydro Québec in 1992. The governor of New York was persuaded that electric utilities in his state could save the equivalent power they planned to purchase from Quebec through relatively low-cost investments in conservation and demand management. Consolidated Edison, for example, a major utility that supplies New York City, will save over $1 billion annually for the next 17 years by reducing demand by 21.8 percent. Under the terms of the Québec Hydro contract, it would have spent $6.5 billion to add 482 megawatts to its power grid. By investing in conservation instead, it will spend $4.1 billion and reduce its need by 2,425 megawatts.

It is because of new ways of thinking like this – named "soft path" economics by renowned energy analyst, Amory Lovins – that megaprojects like Great Whale are being reconsidered and, in some cases abandoned, throughout North America and other parts of the world. Conceived in the 1960s when huge centralized engineering projects were seen as the way of the future, the James Bay project is now being viewed in light of its unaccounted-for costs to native people and the environment. It also appears to be a poor financial investment. As with nuclear reactors, the high construction and maintenance costs and the costs of construction overruns on projects like James Bay, must be born over many years by taxpayers and users of electricity. This is a drain on the Quebec provincial economy and a burden to its economic competitiveness.

Reading 1

Questions and Concerns

READING 1

1. As you are reading this article consider the "gap of credibility" between what was promised by the government and the electrical utility and what actually happened to the native peoples and the environment of the James Bay region.

2. The Cree people are concerned about "the impermanence of the modern economy." What do they mean by this? What proof do they have?

'the geese have lost their way'

THE JAMES BAY HYDROELECTRIC PROJECT HAS TURNED THE LIVES OF NORTHERN QUEBEC'S NATIVES UPSIDE DOWN

By Jamie Linton

You come to the first sign of the James Bay hydro-electric project 251 kilometres north of Val D'Or, Quebec on Highway 109. There is a bridge spanning a deep valley where the Eastmain River used to pass on its way to James Bay. Formerly one of Canada's principal rivers, the Eastmain is now a carcass. Of the 909 square metres per second of water that once flowed through its course, 90 percent has been diverted north by dams and dikes into the La Grande watershed to permit maximum generation of electricity. The roar of the river has been extinguished. Only a pathetic trickle now dribbles through a dry bed.

In his 1985 book, *Power of the North*, Premier Robert Bourassa neatly articulated the vision that is turning northern Quebec into a quilt of generating stations stitched together by high-voltage transmission wire: "Quebec is a vast hydro-electric plant in the bud, and every day, millions of potential kilowatt hours flow downhill and out to sea. What a waste."

According to this view, there is no value in letting a river run its course without harnessing it to generate electricity. So the Eastmain is no longer a waste. By a mighty feat of human engineering, it has been transfused into the La Grande, where it mingles with the water of six other rivers which have been likewise sacrificed. Bearing twice its natural volume of water, the La Grande is disciplined by 215 dams and dikes to flow precisely through a series of penstocks, turbines, and draft tubes. The result is 10,000 megawatts of electricity — almost three times as much as Ontario's Darlington nuclear generator will

Courtesy of Jamie Linton

produce when completed.

Hydro Quebec is the $34 billion, publicly owned utility that has been busy transforming Robert Bourassa's vision into reality. It is its business to reduce the primordial energy of the river to the more mundane form of power that pops our toasters. Like any business, it takes pride in its accomplishments. "Follow the Energy Road!" exhorts one of the Hydro Quebec brochures. "You will experience the infinite landscapes and brilliant skies where thousands of Quebec workers built the La Grande complex."

The "energy road" leads to the town of Radisson on the south bank of the La Grande River, 220 kilometres north of where the Eastmain used to be. Named after the explorer and fur trader who first opened the region for business in the mid-1600s, Radisson is a company town par excellence. Established in the early 1970s to house workers and Hydro officials from southern Quebec, its population of 2,500 still includes only a few native Cree Indians.

For about twice the income they would have received in the south, the workers of Radisson have built a colossal monument. Hydro Quebec's public relations office has three buses to keep up with the 11,000 people who visit the main site each year. Guides trot out an impressive array of statistics: the dam at the La Grande 2 site is the height of a 53-storey building; it contains enough material (23 million cubic metres) to cover the four-lane highway between Montreal and Quebec to a height of seven metres; the tiered spillway cut into solid rock behind the dam is three times the height of Niagara Falls and capable of accomodating twice the flow of the St. Lawrence River; and so on.

The jewel of the piece is the La Grande 2 powerhouse, the largest underground generating station in the world. It is an immense cavern, more than twice the size of Notre Dame Cathedral, cut into the Canadian Shield 137 metres below the surface. Bathed in fluorescent light and throbbing with the revolution of 16 1000-tonne turbines, it is here where the voice of the Eastmain River can still be heard.

Hydro Quebec claims that the environmental and social impacts of taming the rivers are being minimized to the greatest extent possible. "We are dedicated," says its latest development plan, "to developing our hydroelectric resources while mitigating the impact on the natural and human environment." Visitors are told that every consideration has been made for the environment and for the Cree and Inuit inhabitants of the region. There is mention of over 100 environmental impact studies, the expenditure of more than $250 million on remedial measures such as the planting of 10 million trees, and a commitment to ensuring permanent jobs for natives.

This sensitivity to the negative side of development, and the commitment to mitigate major problems, appears to leave little room for criticism. A little homily printed on the placemats of Radisson's main restaurant captures the spirit of the place rather nicely:

"Les monuments sont érigés à la memoire de ceux qui ce sont fait critiquer et non à la memoire de ceux qui ont critiqué."*

If you travel a little further down the energy road, however, a different picture emerges as you approach the Cree town of Chisasibi at the mouth of the La Grande. There was no road and no Chisasibi until the hydro project

---

*Monuments are erected to the memory of those who attract criticism, and not to the memory of those who criticize.

came. Before 1981, the village was called Fort George and it was located on an island at the mouth of the river. The hydro project brought two momentous changes: with a doubling of the La Grande's natural flow, the village was washed into James Bay, forcing a relocation to the south bank. At the same time, construction of the highway instantly plunged the people, who had been living mostly off the land, into the 20th century.

Today, the effects of that radical upheaval are reflected by a roadblock we encounter on the way. A few miles outside of town, the Cree are stopping every vehicle to search for alcohol. You ask about the hydro project and one of the men points to the forest: "See how beautiful it is? It used to be like that (at Fort George) before they built the dams. Now it's all under water. A hunter can go into the woods and stay by a tree so it keeps him warm, just like his mother. We're losing our mother."

Robert Bourassa announced the James Bay project in 1971 without even notifying the Cree. After four years of bargaining and losing a legal contest to stop construction, they and the Inuit signed the James Bay and Northern Quebec Agreement. By its terms, they gave up certain land claims in exchange for special hunting, trapping, and fishing rights, a large measure of self-government, and $225 million to be paid over a 25-year period.

Chisasibi is the most visible result of the agreement. The town has all the amenities of communities found in the south including modern water, sewage, and electrical systems, and an arena, motel, shopping mall, and community centre. Every bathroom wall in the motel sports a Hydro Quebec sticker that says "Save Energy — Use Less Hot Water."

But the new houses and neat subdivision-like appearance belie a harsh reality. Apart from hunting and trapping, which provide a living for about 41 percent of its 3000 people, there is no economic base to support Chisasibi's growing population. Despite the thousands of jobs created by the hydroelectric project, only five residents work for Hydro Quebec. Half the town is unemployed, and the people suffer from alarming rates of alcohol abuse, teenage pregnancy, divorce, and suicide.

These problems were relatively unknown to the Cree before the road was constructed in the mid-1970s. With no warning and no means of coping, the community was suddenly confronted by alcohol, drugs, and the afflictions of modern society. But Chisasibi has begun to fight back. The highway roadblock was established last year, and a drug and alcohol rehabilitation clinic is slated to be opened soon. However, the root of the problem remains: the people are being forced to undergo prodigious change at an impossible pace. It is as if the TV-watching, car-driving, supermarket-shopping inhabitants of our cities were compelled to quit their jobs, demolish their apartments and houses, and go into the bush to fend for themselves.

Because the Cree have been so closely wedded to the land, the James Bay project affects them in a way that is hard to imagine. Northern Quebec's rivers are the foundation of the traditional Cree economy. By altering them and flooding land to create huge reservoirs totalling half the surface area of Lake Ontario, the Quebec government has turned the lives of the Cree upside down.

Even the behaviour of the animals has changed. "The geese have lost their way," laments one old-timer. The caribou seem to have been disoriented too. When 10,000 of them drowned in the Caniapiscau River in 1984, many environmentalists and natives pointed to the alteration of the river as the cause. And as the animals lose their way, the people who depend on them have been quick to follow.

Perhaps the single most devastating effect of the project has been mercury contamination of the fish. In the early 1980s, when two-thirds of the community began showing unacceptably high levels of mercury in their bodies, it came as a shock. Scientists subsequently found that certain bacteria released by flooding the land convert harmless, naturally occurring mercury to poisonous methyl mercury. This is concentrated in the food chain and ingested by people who eat fish from the contaminated waters.

Traditionally, fishing provided not only employment and self-esteem, but a dietary staple for natives. Today, the people are warned by the provincial government to reduce their consumption. Women of childbearing age have been told not to eat fish at all, and no fish caught in the main reservoir of the La Grande and a major portion of the river downstream may be eaten.

In the town mall, mothers leave the food store clutching plastic bags full of processed food. The best-selling item is potato chips; it occupies the entire top shelf of two rows. Surrounded by the bustle of shoppers, the old men gather at a large table. Most of them speak only Cree, but one who introduces himself as David Pepabano finds the words in English: "My children have never seen my world. I have lived off the land, but they don't know anything about that."

He is worried about the future. "Nobody knows what will happen, except Mother Nature. She will fight back. Go watch the TV tonight. You'll see her: big floods, tornados, troubles. I'm happy for that. Mother Nature's fighting back. That's all I'm going to tell you."

The Inuit call it Kuujjuarapik (big, little river); to the Cree, it is known as Whapmagoostui, or "place where there are whales"; the French call it Poste de la Baleine (whaling station); and to the English, it is known as Great Whale. By any name, it is a place that has undergone extraordinary change in a short period of time, with the threat of much worse to come.

Located 200 kilometres north of Chisasibi where the Great Whale River flows into Hudson Bay, Kuujjuarapik is the new frontier. Plans call for the next phase of the James Bay project to be built on the Great Whale River: 80 percent of the river will become reservoir; four dams and 29 dikes will be built, and two more rivers will be diverted to produce 2,890 megawatts of electricity. In order to accomplish this by 1997, preliminary work has already begun on two roads, three workcamp-villages, and three airports in the region. Kuujjuarapik, which was once accessible only by canoe or dog sled, and more recently by plane, will be linked to the south by a paved highway.

While Hydro Quebec maintains that provision for this project was made in the 1975 James Bay and Northern Quebec Agreement, the Cree and the Inuit of Kuujjuarapik are opposed. They have seen the effect the project has had on the environment and communities like Chisasibi, and they want it stopped, or at least delayed in order to give them time to catch their breath.

Matthew Mukash is a spokesman for the Grand Council of the Crees, who have stopped negotiating with Hydro Quebec on the grounds that the utility company has broken its previous agreement. He says,

"Our understanding was that the agreement would allow us to develop at our own pace and to approve any future developments — but that just hasn't worked. The Cree are not opposed to development, but we want to be able to control it in a way that does not threaten our survival as a people."

The Cree are especially concerned about the impermanence of the modern economy. One of their elders, John Petagumskum, puts it very clearly. The 70-year-old trapper had just spent the whole day tending his trapline before being interviewed in the early evening. Through an interpreter, he says that all afternoon the snow had swirled around him, blown up by the wind. "The hydro project will be like that. After it is finished, things will settle and be the same. Nobody will know where, but the money will be gone. So will the fish and the animals."

The Inuit have already suffered considerably. The mayor of Kuujjuarapik, Sappa Fleming, says that the money received from the James Bay Agreement has been spent "like water through our fingers." Meanwhile, his community is losing the ability to provide for itself.

In the past, Inuit hunters shared every bit of meat with the whole community. Through the Hunter Support Program established under the James Bay Agreement, they now receive cash for what they bring back. The meat is distributed to those in the community who need it most, but the motivation for hunting is financial gain instead of community welfare. Hunters are also discouraged from staying on the land any longer than necessary because they are able to collect payment as soon as they bring in game.

Because of disagreements about such changes in their traditional lifestyle, in an agonizing decision made in 1983, the Inuit community split. Half moved to the village of Umiujak, 160 kilometres north, in an attempt to escape the pending devastation. Those who remained in Kuujjuarapik are equivocal about the project. Some believe the only profitable course of action is to negotiate with Hydro Quebec while others, like Mina Weetaltuk, are fighting it outright.

There is a pot of seal meat boiling on the stove in Mina's house. A few goose feathers bound with a piece of string are used to sweep the floor of her kitchen. She is over 70 and speaks only Inuktitut, but Mina wants people to know what she thinks of the hydro project.

Through her niece, she says she fell on her back ice fishing a week before, and wasn't feeling as perky as usual. But you'd never know it: "I have seen those ribbon-cutting ceremonies on television," she declares, her face radiant with energy, "and if they try it here, I'll go and sew the ribbons back together. If they try to put their dams on our river, I'll beat them back with a wooden stick."

Adopting a more subtle strategy last April, Mina and 60 other Inuit and Cree Indians went by dog sled from Kuujjuarapik to Chisasibi, then flew to Ottawa, where they began paddling to New York City to protest the next phase of the James Bay project. The purpose of the trip was to alert Americans to their cause, as Hydro Quebec wants to harness the Great Whale River to generate electricity for the US market. Designed and constructed by Mina's husband, Billy, the vessel they used was half canoe and half kayak, symbolizing the Cree-Inuit solidarity against the project.

Many of the people had never been "south" before, and for the first time in her life, Mina was unable to maintain the "country" diet of meat and fish she is accustomed to. Having to stomach the likes of hamburgers and Coke put her in an Ottawa hospital for four days. But she survived the ordeal and rejoined her friends for the voyage to New York. They made it in time for the Earth Day celebration in Manhattan.

In the municipal office of Kuujjuarapik, there is a small glass case for pictures and handicrafts. On the bottom shelf is a miniature, hand-carved wooden paddle commemorating that journey. Inscribed on it is the message: "To the Inuit of Kuujjuarapik from the Cree nation of Whapmagoostui, in commemoration of joint efforts to save our homelands and our way of life from destruction by the hydroelectric project."

The seal meat on Mina Weetaltuk's stove is so pungent it almost knocks you out. It is the stuff from which she gets her remarkable energy — the same energy that feeds John Petagumskum's wisdom and animates the spirit of the Inuit and the Cree of Kuujjuarapik in their resistance to the next phase of the James Bay project. And it is the same energy that is felt in the presence of a wild river. It seems an unforgivable crime to extinguish such a force to produce more of the brute power on which we have grown so dependent. ❖

Danny Beaton, a band member of the Mohawk Nation, assisted in the research of this article.

Reading 2

Questions and Concerns

READING 2

1. As you read the letters from Ambassador Burney, Chairman Drouin, and the National Resources Defense Council try to read between the lines. Underneath all three written statements are political and philosophical agendas. (It is most interesting to re-read the first two letters after reading the rebuttal made by the NRDC.)

JAMES BAY: OPEN LETTERS TO MEMBERS OF NRDC

The James Bay wilderness in northern Québec is the site of one of the largest and most destructive energy projects in North America: a series of dams that have flooded 4,000 square miles of forest, altering wildlife habitat and the way of life of the native local peoples. Hydro-Québec, the utility that built the project, is planning a second phase that would begin on the Great Whale River. If this second phase is completed, the additional reservoirs created would ultimately cover 4,000 square miles.

NRDC has been working with the Cree Indians who live in the region, and other concerned Canadians, to oppose the project and promote energy efficiency. Last spring, NRDC Executive Director John Adams and Matthew Coon-Come, Grand Chief of the Cree, wrote NRDC members to tell them about James Bay. John Adams asked the members of NRDC to write to Hydro-Québec and the Canadian embassy and express their opposition to the project. Over ten thousand did so — an overwhelming response that immediately caught the attention of both the utility and the embassy.

In the interest of furthering dialogue on the issues, NRDC invited the Canadian Ambassador and Hydro-Québec to address our members directly in these pages. We have included a response from the three NRDC staff members who have taken the lead on the project.

Look for updates in future issues of The Amicus Journal.

Letter from the Canadian Ambassador

501 Pennsylvania Ave., N.W.
Washington, D.C. 20001

On behalf of the Canadian government, I welcome the opportunity to clarify the position of the Canadian government on the proposed Great Whale project.

In 1975, the Canadian government signed the James Bay and Northern Québec Agreement (JBNQA) with the Québec government, representatives of the Cree and Inuit, and Hydro-Québec. The Cree and Inuit were signatories to the agreement, which explicitly recognized their rights and governs future development in the region — such as the proposed Great Whale project — only after a thorough environmental impact assessment.

The Cree and Inuit are full participants and equal partners in the five environmental impact assessments being jointly conducted by governments in Canada on the proposed Great Whale project. The governments of Canada and Québec have declared respectively that no decision will be made and no construction will begin until the reviews are completed. As outlined in the box, the guidelines for the environmental impact statement (EIS) are amongst the most comprehensive ever issued for such a project in Canada.

Courtesy of The Amicus Journal (Winter 1993)

The JBNQA provides the 18,000 Cree and Inuit of Québec a degree of autonomy and self-government unequaled elsewhere on the continent. Among the most critical changes have been the provision of modern services in education, health care, communications, and justice — services which are now administered and controlled by the Cree Regional Authority. Cree population has doubled since the beginning of the James Bay project in 1975. Infant mortality has decreased by one-half and life expectancy has increased from between forty and fifty years of age to about seventy, which is comparable to the national average.

To help the Cree preserve their traditional way of life, the Québec government provides an income security program, which subsidizes traditional activities. In an interview with Montréal's daily *Le Devoir* published on September 23, 1991, Cree Grand Chief Matthew Coon-Come said that this program "has strengthened our economy as hunters, fishermen and trappers....This success story makes us the envy for all the Natives across Canada and even internationally." With funds received from the JBNQA, the Cree and Inuit have also begun several nontraditional businesses. These include construction companies, two airlines, a boat manufacturing firm, and services such as restaurants and motels.

Much has been said about the drowning of 10,000 caribou in September 1984. Studies by the Canadian and Québec governments have concluded that hydro development was not the cause; in fact, the upstream facilities lowered the Caniapiscau River's flow. Over the last twenty years, the caribou population has increased from 200,000 to 700,000.

Any concerns about the potential environmental effects identified in the EIS and the proposed Great Whale project can be addressed frankly and openly through the public review process. We welcome input to this process, particularly when it is constructive and based on consideration of all the facts.

D. H. Burney
Ambassador

Letter from Hydro-Québec

75, boulevard René – Lévesque ouest
Montréal (Québec) H2Z 1A4

We welcome this opportunity to address the members of the NRDC. Hydro-Québec is committed to supplying electricity to the people of Québec and to developing the hydroelectric potential of the province. To do so, Hydro-Québec is also committed to protecting the environment and respecting the rights and aspirations of the Native people of Québec, including the Crees and the Inuit.

Unfortunately, the debate over the proposed Great Whale project has left an inaccurate and, we believe, unfair impression of Hydro-Québec with the environmental community in the United States. We would like to set the record straight.

That record is clear: Hydro-Québec is committed to conservation and demand side management (DSM) programs and it is dedicated to preserving the environment as it develops Québec's energy resources. *[Editor's note: Demand-side management programs seek to control the "demand" side of the energy equation (through energy efficiency and other means), rather than to increase the "supply" side by generating more power.]*

Conservation at Hydro-Québec

Conservation has been a priority at Hydro-Québec for thirty years, due in part to climate and geography. Electrical demand in Québec is nearly double in the winter than during the summer. That means Hydro-Québec must tailor its DSM programs to reduce winter demand.

Through measures initiated by Hydro-Québec, the Province of Québec and the Government of Canada since 1963, Québec homes are among the most energy efficient in the world. Our programs have included improved electrical wiring, energy-efficient water heaters and insulation standards for building construction. In 1983, Hydro-Québec's home insulation standards developed over two decades were written into Québec law — the first law on energy savings in buildings enacted by a Canadian province.

Compare Québec to the Province of Ontario. While the weather is warmer in Ontario, its average electric consumption for heating a single family home is 14 percent higher than Québec. The difference is even greater when compared with the northeastern United States, where the climate is similar to Ontario. In fact, Hydro-Québec is recognized as a leader in North America in peak load management through its residential, industrial dual-energy, and interruptible power systems.

Moreover, a survey of United States utilities revealed that 0.7 percent of total electric revenues are spent on demand side management. In contrast, Hydro-Québec spends 3 percent of its revenues on DSM. By the year 2000, our conservation plans should produce annual savings of 9.3 billion kilowatt-hours, roughly equivalent to New York State's electric needs for twenty-three days.

Why Hydropower?

The majority of the people of Québec believe in hydropower because it is a clean, safe, renewable and inexpensive source of energy. From our perspective it is clearly superior to the alternatives.

That is why 95 percent of our electricity needs now come from hydropower. As a result, our greenhouse gas and acid rain emissions are relatively low. Indeed, per capita emissions from energy sources are less than half as high in Québec as in the United States.

Moreover, hydropower is more efficient than fossil-fuel energy. Homes heated with hydropower-generated electricity are 90 percent energy efficient, compared to 60 percent for homes heated with oil.

Great Whale

Hydro-Québec proposed building the Great Whale project to meet the energy needs of the people of Québec. We believe conservation is important, but even if our DSM and conservation programs are successful, Québec will still need new electric sources in the future.

The 1975 James Bay and Northern Québec Agreement, which was approved by, among others, the Crees and the Inuit, the Government of Canada and the Province of Québec, recognized Hydro-Québec's right to undertake the Great Whale project providing that it complies with a rigorous environmental review process.

We are now preparing an environmental impact statement for the Great Whale project based on guidelines issued last month by the five environmental review panels overseeing our EIS process. These guidelines are unprecedented in their scope. They require significant amounts of information and analysis on the environmental, economic, social and cultural impact of our project. Hydro-Québec has been studying Northern Québec and the Great Whale region for 20 years and we will comply with the guidelines.

Furthermore, we won't undertake any construction before obtaining all the necessary authorizations from the Canadian and Québec governments.

Mercury

There is no doubt that our hydroelectric projects have an impact on the environment. But the allegations by opponents too often overstate the risk and understate our efforts to mitigate these impacts. Mercury is a good example.

Creation of the reservoirs for the La Grande projects led to increased levels of mercury in fish, which is a major part of the diet of the Crees.

Mercury levels in fish in Northern Canada are naturally high, due in part to industrial and power plant pollution from the United States and Canada. But when Hydro-Québec discovered higher-than-normal mercury levels in the early 1980s, studies began and an agreement was signed with the Crees that established a mercury monitoring program on the possible health, social and economic impact on the local population. Moreover, research revealed that mercury levels diminish over time and return to their original levels after 20 to 30 years.

Since the inception of the monitoring program, the concentration of the mercury levels measured among the Crees has decreased. The Cree Board of Health concluded in the 1989-1990 Mercury Committee Annual Report: "An examination of the results of the past few years confirms that the majority of Crees have a level of mercury exposure that not only is not problematic for their health, but also permits a promoting of the nutritional value of fish in their diet."

Hydro-Québec is proud of its record for producing energy for Québec and we are willing to discuss it with anyone. But misinformation, exaggeration and distortion do not contribute to a fair debate and can only obscure the very real and serious issues Québec faces as it confronts its energy needs for the next century. We welcome a discussion of the real issues and we would be pleased to provide additional information to any member of NRDC who requests it. You can send your requests to:

Hydro-Québec, C.P. 6071
Montréal (Québec), Canada H3C 3A7

Richard Drouin
Chairman of the Board
and Chief Executive Officer

Letter from NRDC Staff

Since we first wrote to you there has been significant progress in the environmental assessment process for the Great Whale Project, due in large part to litigation by the Cree and international pressure — including the many letters and postcards sent to Hydro-Québec and the Canadian Embassy by NRDC members. As presently outlined, this review *could* be among the most comprehensive environmental reviews that have been undertaken for hydro development projects. But it remains to be seen how Hydro-Québec will respond to its obligations under the new guidelines. It is important that international pressure be maintained to ensure that there is full and meaningful public participation as the review process unfolds.

As to the letters submitted by Canadian Ambassador D.H. Burney and Hydro-Québec's Richard Drouin, we offer these comments:

• Contrary to suggestions in both letters, the full environmental review is not the product of good will by Hydro-Québec, Québec, or Canada. It is the result of numerous lawsuits. Even as they are participating in the review process, Québec and Canada are simultaneously trying to have the decision that requires federal environmental review reversed in court. Further, Hydro-Québec has at least four other large hydro projects proceeding at present without any public participation in the review process.

• Ambassador Burney argues that the Cree and Inuit were signatories to, and received compensation from, the 1975 James Bay Northern Québec Agreement (JBNQA) that allows hydro development on their homelands. Ambassador Burney nowhere mentions that the Cree signed this treaty under duress, after Hydro-Québec had already entered their lands without their permission and begun construction.

• Ambassador Burney points out that the JBNQA agreement provides the Cree with compensation and with some social programs. However, many of these programs have been provided to other Canadian citizens for years without any quid-pro-quo agreement. Furthermore, according to the James Bay Cree, many of the promised benefits either have not been delivered or have been delivered improperly.

• Mr. Drouin's letter discusses Hydro-Québec's commitment to energy conservation and demand-side management programs. NRDC's review of Hydro-Québec's energy efficiency efforts shows that, when compared to successful utility programs, Hydro-Québec's programs are poorly designed and underfunded. They will therefore capture only a fraction of the energy savings that are possible. Properly designed efficiency programs would eliminate the need for new supply, cost less than the Great Whale project, and create more long-term jobs.

• Both Ambassador Burney and Hydro-Québec claim that Hydro-Québec's spilling from the Caniapiscau Reservoir did not cause the drowning of 10,000 caribou downstream. However, the Government Secretariat for Indian Affairs, the agency that investigated the event, found that "the management of the Caniapiscau Reservoir was the principal cause of the drownings." Nor is there any record or memory of a drowning of this magnitude before the dams.

• Richard Drouin maintains that the problem of high mercury levels "is not problematic" for a majority of Cree and that Hydro-Québec's research shows that the mercury concentration in fish will diminish "after twenty to thirty years." In fact, two-thirds of the Cree in areas affected by the first phase of hydro development in the James Bay region suffered from mercury exposure that exceeded World Health Organization standards. To the extent that these levels have dropped, it is because the Cree have stopped eating fish, their traditional food and one of the important bases of their culture, in favor of canned or frozen food. Health advisories and prohibitions on fish consumption remain in force in the affected areas. Furthermore, federally funded studies by Winnipeg's highly regarded Freshwater Institute concluded that mercury levels in fish will not normalize for eighty to ninety years.

We look forward to continuing our discussion with the Canadian Embassy and Hydro-Québec. We are confident that if the environmental review is performed properly, it will show that the second phase is unnecessary and should not be built. We will continue to participate in the review process and to work on this most important issue with our friends in Québec and the members of NRDC.

Ashok Gupta, Senior Analyst
Robert F. Kennedy, Jr., Senior Attorney
S. Jacob Scherr, International
Program Coordinator

Questions and Concerns

READING 3

1. Read this article then compare the authors' conclusions with the statements made by the chairman of Hydro Québec in Reading #2.

2. If the authors are correct, why would an electric utility like Hydro Québec not commit fully to demand management and conservation programs that would save electricity?

MEGAWATTS OR NEGAWATTS?

IN THE CAMPAIGN TO PROTECT THEIR LAND FROM FURTHER development, the Crees and Inuit of Quebec are trying to convince fellow Quebeckers and residents of the northeastern United States to look to their own homes and workplaces to satisfy their appetite for electric power, rather than to the wilds of central Quebec. It is a persuasive argument: a raft of detailed analyses as well as lessons gained from innovative utilities suggests that a comprehensive energy-conservation programme could neutralize the need for the Great Whale hydroelectric project.

Efficiency technologies developed over the past decade have dramatically altered the energy equation. Energy analyst and longtime conservation proponent Amory Lovins of the Rocky Mountain Institute in Colorado estimates that conservation measures now available – from old-fashioned weatherization to advanced lighting – could save between 30 and 70 percent of total energy consumption at a cost below that of building new power plants.

Utilities are taking heed. In Canada, B.C. Hydro pioneered the energy-efficient route with its Power Smart programme. Power Smart offers comprehensive efficiency measures to consumers and businesses such as rebates for every kilowatt saved through the use of high-efficiency motors and water heaters, as well as loans and rebates to install efficient lights and to upgrade insulation in homes. By the year 2000, Power Smart efficiencies are expected to save the amount of power now consumed by 400,000 homes in the province. Consolidated Edison, a New York utility that will claim almost half of a proposed 1,000-megawatt power contract between Hydro-Québec and the New York Power Authority, is planning to save billions of dollars over the next 17 years by reducing electricity demand by 21.8 percent. The analysis is telling: By investing in Quebec hydroelectric power, Consolidated Edison will pay $6.5 billion to add 482 megawatts to its power grid; but by investing in conservation, it will pay $4.1 billion to reduce 2,425 megawatts from its capacity needs. The savings come from the electricity *not* produced, something Lovins calls "negawatts."

Quebec itself is a prime candidate for a multi-pronged energy-conservation programme. About one third of Quebec's nonapartment housing stock has little or no insulation, according to a recent study conducted for the department of Energy, Mines and Resources. Residential energy-conservation programmes designed to upgrade houses to present standards could save 50 percent of the energy production of the Great Whale project. The potential for savings from the commercial sector is also substantial. Concordia University's Centre for Building Studies estimates that the typical office building in Montreal consumes 30 percent more energy than comparable buildings in northeastern United States. "Our figures show that an aggressive conservation programme could delay the need for the Great Whale project until at least 2010," says Ian Goodman of The Goodman Group, a Boston-based consulting firm. "By that time, technological advances may well have supplanted the need for the massive dam complex."

For many in Quebec, however, the Great Whale project has merits extending beyond those of energy generation. They see the project as an economic engine, creating jobs and stimulating the economy into the next century. But research into energy conservation questions the prevailing wisdom. Work by economists Mark Jaccard and David Sims for B.C. Hydro reveals that twice as many jobs would be created through energy conservation without the creation of a new dam and that the jobs would be more evenly spread geographically.

Far from securing Quebec's economic future, says Goodman, the construction of the Great Whale project may saddle the province with an energy glut as American utilities increasingly look to conservation. As a sign of the times, Vermont's largest utility in 1991 had to sell 70 megawatts of newly acquired Hydro-Québec power back to Quebec. "The demand just isn't there," says Lewis Milford of the Conservation Law Foundation in Montpelier, Vermont. Without the projected income from energy sales to pay down the huge debt burden of the James Bay megaprojects, Quebec consumers will face higher electric bills and their economy's competitiveness will suffer. In the end, ironically, scrapping the Great Whale project may offer a more secure future not only for the aboriginal peoples in central and northern Quebec but for the province as well.

— *Pegi Dover and Philip Jessup*

Courtesy of Pegi Dover and Philip Jessup

Reading 4

Questions and Concerns

READING 4

1. Why would the government of Québec agree to provide electricity cheaply to large corporations at a loss?

Secret pacts cost billions, Hydro-Québec studies show
Liberals accused of hiding facts from average consumer

BY RHÉAL SÉGUIN
Quebec Bureau

QUEBEC — Hydro-Québec's secret risk-sharing contracts with aluminum and magnesium producers in the province will result in billions of dollars in lost revenue, according to the Crown corporation's own studies completed last January.

The government has refused to release the studies but members of the opposition have obtained copies. Yesterday, the opposition politicians accused the ruling Liberals of hiding the facts from Quebeckers and forcing the average consumer, through higher electricity rates, to pay for what amounts to a corporate giveaway worth hundreds of millions of dollars a year.

"Can we afford to lose billions of dollars over the life of these contracts? The government is asking Quebeckers to tighten their belts. Can we afford it?" Equality Party member Robert Libman asked yesterday in the National Assembly.

Parti Québécois energy critic Guy Chevrette demanded that the government officially release the studies. "Will the Energy Minister confirm that Hydro-Québec has studies for the duration of the contracts until the year 2010 ...and that the balance sheet shows they will end in a fiasco, a $2-billion loss for all Quebeckers?" Mr. Chevrette asked.

The contracts, which took effect in 1987, give aluminum and magnesium producers preferential tariffs based on world market prices of the metals produced. Premier Robert Bourassa's government, responsible for getting Hydro-Québec into the arrangement in an attempt to attract aluminum and magnesium producers, has long argued that the deal would generate enormous profits for the Crown corporation.

Aluminum and magnesium prices are not expected to meet government pro-

> **'In a freer market economy, the Great Whale River project would be scrapped.'**

jections and it is estimated Hydro-Québec will lose $2.9-billion over the life of the 23-year deal, the studies indicate.

The Hydro-Québec studies pretty well confirm what Northern Quebec Crees have maintained. In 1991, a study by Robert McCullough, a U.S. consultant for the Crees, suggested that total losses under the contracts could reach $2.3-billion.

Hydro-Québec's fixation on subsidies to the metals industry, the utility's loss of export contracts and its huge debt incurred in the pursuit of megaprojects is breaking the Quebec economy, Diom Saganash, deputy grand chief of the Northern Quebec Crees, said in an interview recently. "It is also what leads to the destruction of Cree lands and our way of life...In a freer market economy, the Great Whale River project would be scrapped."

The figures, originally leaked to the French-language television network Société Radio-Canada, prompted angry exchanges in the National Assembly yesterday. Energy Minister Lise Bacon accused the opposition of attempting to foil Quebec's attempts to attract new investors.

Ms. Bacon has argued in the past that publicizing details of the contracts would allow supporters of protectionist policies in the United States to protest that Quebec industry was getting subsidized electricity. Quebec obtained a court injunction blocking the release of the details of a deal signed with magnesium producer Norsk-Hydro. While that deal has since been renegotiated, Hydro-Québec's recent projections still that the Norsk-Hydro contract will spell a loss to the utility of $200-million over the life of the contract.

Ms. Bacon argued yesterday that the investments made possible by the risk-sharing contracts created 8,000 jobs and generated $600-million in salaries. She said the economic spin-off of the investments amounts to $1.5-billion over the life of the contracts.

She did not deny that none of the 13 shared-risk contracts will produce a profit for Hydro-Québec.

Courtesy of The Globe and Mail (March 31, 1993)

Discussion for The James Bay Hydroelectric Project

1. Most people believe that huge engineering projects like the James Bay hydroelectric complex or the Darlington nuclear power station in Ontario are testaments to "progress;" that if our society is to move forward and grow, we need such installations; and that there are no realistic alternatives. But others say large, expensive electricty-generating megaprojects seldom pay their way; that they are extremely expensive; and they need not be built if we simply adjust the ways in which we use electricity. Discuss your concept of progress. Does it always mean "big" and "growth-oriented"? Isn't progress better defined as a solution that generates the greatest good at the least cost while harming the fewest people and preserving the environment?

2. Another worthy point of discussion is the centralization of large power stations and dams. Some people argue that many smaller, local stations would be preferable (and likely less expensive) than a single massive hydroelectric or nuclear project. The impact on people and the environment would be decreased, and alternative and relatively clean-burning fuels like alcohol (an agricultural byproduct) could be burned to generate electricity. That way, smaller stations could go in and out of operation as needed without affecting the power grid, and more jobs would be created over the long term within many communities. Do you think this is realistic? Discuss the concept of spreading the risks and benefits of a large project (like a power station or even a large centralized company) over many smaller units. How is this good for the environment? How is it better for people?

3. If the natives of James Bay succeed in preventing construction of the second phase of the hydro project, a few thousand people living in a remote part of the Canadian wilderness will have changed the plans of a major provincial utility that has a responsibility to supply electricity to millions of people. Is this right?

4. The concepts of "demand management" and energy conservation are critical to understanding the arguments for and against the James Bay project. Proponents of energy conservation show that a kilowatt saved is one less that has be produced, so that by investing in the management of demand rather than endlessly meeting a growing need for supply, electric utilities can avoid the costs – financial, environmental, and social – of building new power stations. Environmentalists argue that the years of time we gain through this strategy should be used to research more benign forms of energy, such as solar, wind, geothermal, biomass, and hydrogen. Do you think the future rests in big megaprojects, or should they be supplanted – by conservation in the short run and less-polluting energy sources in the long run? Should science focus on making better dams and nuclear reactors or on finding alternatives that are less intrusive on our society and our environment?

5. On top of incurring massive debts to build the first phase of the James Bay project, it appears from Reading #4 that Hydro Québec also signed secret long-term contracts with preferred industrial clients in the magnesium and aluminum businesses. These will cost the taxpayers of Québec further billions in lost revenue. Consider the statement by the deputy Grand Chief of the Northern Quebec Crees: "In a freer market economy, the Great Whale River project would be scrapped." Why? What does this say about monopolistic public organizations like Hydro Québec?

Introduction

The Built Environment

This issue examines the environmental efficiency of houses and other buildings, and how Canada is a world leader in the development of environmentally progressive housing.

Although most of us don't think too much about construction of our "built environment" – homes, offices, institutions, factories – buildings have a major effect on the environment. In aggregate, they are one of the largest contributors to global warming, next to the automobile the largest contributor to ozone depletion (through CFC emissions from air conditioning and refrigeration), and a major draw on fresh water supplies.

For these reasons, the way we build the structures in which we live and work is coming under greater scrutiny from governments and other organizations committed to energy efficiency, water conservation, and waste reduction.

Think about the lifetime implications of an environmentally *inefficient* building. A building that wastes energy and water can, theoretically, do so for a century. Buildings normally live on well past the lifespans of their occupants, so the qualities enshrined in buildings today, or not enshrined, will have had significant cumulative repercussions by 2092 or 2192. A house that emits an extra ton of carbon dioxide per year because it lacks insulation will emit 100 extra tons over a century. Therefore, environmentally efficient construction is a responsibility to future generations.

Some experts are predicting that by the end of this decade the need to curtail global warming will start to assume the same level of urgency that we now face with ozone depletion. If so, building codes will change in response, and quite possibly expensive retrofits of wasteful buildings will have to occur. The obvious preemptive solution is to construct environmentally efficient buildings in the first place.

The ultimate waste, of course, are structures so poorly designed and assembled that they don't last. Not only are they wasteful during their lives, but they become waste themselves – huge amounts of demolition waste – and have to be replaced with new materials. Entire suburbs and public housing projects in North America built after World War II have already been torn down to make way for new developments because they simply were not worth keeping.

There are signs that the construction industry is improving its methods, materials, and approaches to waste as our collective understanding of the environmental impact of buildings grows:

• In the early 1980s, the Canadian federal government sponsored a research program to set more efficient standards in home heating, ventilation, and insulation technology. Called R-2000, it encouraged homebuilders to construct houses that used half the energy of conventional houses of the same size. The R-2000 standard is now quite common across Canada. The city of Toronto's building code requires it of all new structures.

• In 1992, the federal ministry of Energy, Mines, and Resources in conjunction with the Canadian Homebuilder's Association took the next step forward with the Advanced Houses Program. Eleven model Advanced Houses located in eight provinces improve on the R-2000 standard by at least 50% in space heating, cooling, water heating, appliance draw, and lighting. The models are also partially constructed of recycled and recyclable building materials, use water-saving devices like low-flow toilets to reduce normal water consumption by about half, and use no chlorofluorocarbons (CFCs) except for refrigeration.

• Canada Mortgage and Housing Corporation (CMHC), the federal government's housing authority, sponsored the Healthy Housing Design Competition in 1991. The winner (who was announced in late 1992) had to design a sustainable home that showed a "systems approach," taking into account the global environment, the indoor environment, and affordability. CMHC hopes to encourage innovation in the building industry, including more sustainable building products and services.

• Sellers of used building materials are opening across Canada. At least four (in Winnipeg; Waterloo; Ottawa, and Halifax) are non-profit ventures called "ReStore," affiliated with the affordable housing organization Habitat for Humanity. Others, such as Architectural Clearinghouse in Edmonton; ReUze Building Centre in Toronto; and Hobo Hardware in Guelph are for-profit businesses. They are all seeking to extract reuseable materials from buildings being torn down or renovated. (Almost all the materials in old buildings demolished in North America today with the exception of brick and large beams are sent to landfills.)

• The Government of Ontario is working with Ortech International, a private research firm, to ensure that recycled building materials are used in the construction of new government offices. The program will "change specifications to incorporate the 3 Rs (Reduce, Reuse, Recycle) into the design stage" of development.

• North America's largest wood recycling plant (in Brampton, Ontario) was opened in 1990. In Metropolitan Toronto and some surrounding municipalities it is now unlawful to dispose of wood and drywall in landfills.

• A company in British Columbia has invented a machine that remelts old asphalt. Another Canadian company has devised a way to strip the paper backing off drywall and recycle the gypsum.

• The cities of Kitchener and Waterloo in Ontario are countering a shortage of drinking water in their region by installing low-flush toilets in every home free of charge.

• Indoor air quality is a growing environmental issue. The U.S. Environmental Protection Agency (EPA) has released a number of reports on the hazards posed by second-hand cigarette smoke and chemical outgassing of common office fixtures like carpeting and particle board. There is also speculation that greater numbers of people are suffering from acute environmental sensitivity syndrome, possibly due to their exposure to pollutants. Older heating/ventilation, and air conditioning (HVAC) systems in office buildings often do not provide enough ventilation, recirculating stale air rather than introducing fresh air. Some companies that have modified their HVAC systems have found less sickness and absenteeism among employees and a boost in productivity.

Why aren't all new buildings constructed to be environmentally efficient now that we understand these things? It comes down to the perception of value. Measured in dollars spent today, cheaply built and ecologically inefficient buildings are almost always less expensive to erect. But when measured in ongoing costs (energy, water, repairs), the cost to the environment (greenhouse gases, landfill waste), and the human cost (bad air, a depressing lack of quality), these structures are in fact far too "expensive." Environmentally superior homes, offices, and factories are, in the long run, less expensive and more valuable. They use less energy. They cost less to maintain. They will have higher resale values. And they feel better – people quite naturally want to be associated with progressive, efficient structures that embody environmental forethought.

Reading 1

Questions and Concerns

READING 1

1. The author of this article, Graeme Robinson, is from New Zealand, but his opinions apply to buildings everywhere, especially in a developed northern country like Canada where energy use in buildings is high. He says our "urban systems are a direct manifestation of our life styles. In building them we declare our vision for the future and commit our resources to that vision..." What do you think he means by this?

2. Think about the attributes a "sustainable built environment" should have.

How Green Can The Built Environment Become?

Graeme Robertson

Introduction

This article explores the possible limits in terms of what is practicable and necessary in the production of a more sustainable built environment.

The list of requirements for sustainability has widened considerably during the last decade. The very restricted concern for energy efficiency of the actual building in use has lead to aspects involving the design, construction, use and eventual reuse of the components of the obsolete building, along with the actual placement of that building to allow for a reduction in the need to move people and goods.

It is also increasingly realised that to constrain investigation to just new buildings is not sufficient – the existing building stock will always be the dominant component. We must modify existing buildings as well as design and construct our new buildings in a more sustainable manner.

We know what the necessary steps are. We also know that these same steps are invariably cost effective and produce better, more user-friendly environments.

Increasingly it is apparent that the perceived limits are imposed by economic controls that are essentially conservative and limiting, with short term economic goals dominating at the expense of more responsible concerns for future generations. In a country – indeed a world it seems – driven by economic values, just what possibilities are there for change? Can we expect, in the short term, a significant alteration in the formation and use of our built environment to satisfy the demands of a sustainable society? The attitudes of those who control the economic sector must change significantly (1). Too often proposed environmental changes are seen as a threat rather than an opportunity.

The Need

How important is the built environment contribution to the achieving of a sustainable society?

Human settlements, from tiny hamlets to great metropolitan cities, have been an integral part of our cultural and technological development. Our urban systems are direct manifestations of our life styles. In building them we declare our vision for the future and commit our resources to that vision; in doing so we consolidate and perpetuate our vision into the future.

Courtesy of Graeme Robertson

Faced with new realities, we must re-develop our visions and our life styles. Correspondingly we must redevelop our buildings and our urban systems. Currently our buildings and the infrastructure that supports them represent up to 75% of our accumulated capital stock (2). Together they are the principal life support systems for our future. As such, they commit us to future patterns of behaviour and future levels and patterns of environmental impact. The conversion, and perhaps extension and replacement of this accumulated capital stock must become a significant part of the process through which new and sustainable relationships within the biosphere can be established.

Buildings use energy for lighting, heating, air handling and air conditioning, and they support energy-using equipment (3). The making of buildings consumes energy embodied in products such as steel, aluminum, glass, timber, plaster, bricks and cement (4). Energy is also used in transporting materials and components to and from the building site, and then in the construction process as they are hoisted into position and as tools and equipment are used to fit them.

Every time a material or fitting is chosen there is a decision to use energy and thus to release greenhouse gases. In deciding the overall form or orientation of a building, patterns are established that determine access to solar energy and daylight. Every window is both a source of energy or light and an area through which heat can leak from the building.

Individually these design decisions are insignificant, yet collectively they constitute a major part of the greenhouse problem (5). They express our consumptive industrialised society; they trap us within short life buildings that are larger than we need, have low levels of utilisation, are constructed of energy intensive materials such as aluminum, steel and cement, and which, because of their design, require a continuing supply of energy to achieve suitable air quality, thermal and lighting conditions.

There are other decisions about buildings which establish future environmental impacts. It is decisions about location and function that determine future transport needs. In pursuing our various individual life styles we use a variety of buildings and urban facilities. To do this we move from one to the other. When the distances are greater than can readily be accommodated by walking we use mechanical transport systems of one kind or another. Each has its own level of environmental and social impact. By far the most energy efficient are rail based mass transport systems; least efficient is the single occupant private motor car (6). All, however, cause degradation through consuming non-renewable resources and polluting the environment. All lead to ill health and trauma. All consume that most precious of all our non-renewable resources, time.

The composition and the dimensions of the built environment are thus primary generators of the environmental and human degradation attributable to mechanical movement. A decision to develop at low density and to support this with extensive motor transport is a decision which commits future generations to finding suitable fuel and emitting greenhouse gases.

Because these many different contributions and involvements are normally identified as being separate, their pervasive influence in establishing and then perpetuating patterns of energy consumption is obscured. Together they make the dominant contribution to our environmental prospect.

The Solutions

An essential part of the transformation of the built environment must be the development of an attractive human habitat. A solution cannot be achieved through direction. Rather, it must be presented as an attractive alternative to the present situation such that individuals and communities will want to make the changes that are needed. Thus the new situation must be friendly both to the environment and to those who will make it and use it.

A sustainable and fulfilling built environment system of the future will be based on renewable and environmentally clean energy and materials. As long as hydrocarbons remain the principle source of energy, and in particular of motive power (7), we should be developing a built environment that minimizes the movement of people and goods. In parallel with the development of environmentally friendly new buildings there will be an equally pressing need for a transformation of the existing stock. Such a double pronged approach greatly increases the rate at which we can change the operating characteristics of the future built environment stock.

The opportunity facing us is to participate in the greatest building programme of all time. The prize for this and future generations is to achieve a new physical basis for long term sustainability.

The Hindrances

In Roger Douglas parlance, the playing field is far from level. Just how can politicians and business leaders be convinced to modify their rather established positions and accept the advantages and indeed necessities of a

Necessary actions

- encourage and perhaps demand the construction of buildings that use low levels of non-renewable energy in all phases of their manufacture, construction, use and disposal;

- encourage the development and use of renewable energy in buildings and throughout the built environment system;

- encourage the re-use and recycling of buildings, building components and building materials over long periods of time;

- encourage the development of environment-friendly technologies that we would welcome as neighbours;

- encourage technologies and management techniques that help bring closer together the various components of our life supporting systems;

- encourage the continuing development and widespread availability of computing and telecommunications – the smart society cleverly managing its relationship with its social and physical environment;

- encourage local complementarity of built environment components and facilities leading in time to local self sufficiency – including residential infill in central areas and non-residential infill in existing residential areas;

- encourage the local use of organic wastes for fuels and thus avoiding the release of methane in favour of carbon dioxide;

- encourage local systems for harvesting and recycling water and wastes;

- encourage every available area to be used for productive purpose including energy harvesting and urban food production;

- discourage technologies and management techniques that physically separate the various components of our life support system.

'Green' future? How can short term economic or electoral demands be changed to accommodate considerations regarding the long term viability of life on this planet? Balance sheets or polls of voters do not at present reflect the growing importance of the environmental lobby (particularly in New Zealand). The decision makers still have to be convinced (8).

Accepting that we as a society know what should be done to achieve a sustainable built environment, then the problem is how we instigate change. To debate endlessly with the holders of the short term economic or political viewpoints becomes pointless in the extreme. The most informed in that camp are fully aware of the threats but choose to ignore them. It is not out of ignorance that their opinions are created but out of selfishness and conservativism. It is better to motive the people – the users of our built environment – so that they will demand significant changes by our so-called leaders.

And of course what is being offered as the sustainable alternative is remarkably user friendly and easy to 'sell.' We as users of the built environment do not have to be convinced. We know that buildings that are responsive to the natural environment, by using daylight, solar heating and natural ventilation, are appreciably better. We know that a society based on efficient mass transport systems, or better still only requiring a short walk from home to work etc, is rather nice. We also know that the reuse and recycling of the components and materials of the built environment can produce many other aesthetic benefits beyond those related to the efficient use of non-renewable resources.

It appears that we have the evidence.

It also appears that we have the solutions available.

It certainly appears that the users of the built environment, the people, are ready to be fully convinced that a sustainable future for the built environment is in fact one related to opportunity not negative constraints (9).

The politicians and business leaders will then follow.

End Notes

1. There is a widespread argument that achieving a sustainable built environment means controls and regulations which in turn limit personal freedoms and free-market philosophies which eventually lead to a degradation of our international trading position. Possible performance based regulations in terms of carbon dioxide emission audits and carbon taxes are not anti free market philosophies. They represent a method of informing the market of the true costs involved. Sweden is becoming one of the most environmentally-friendly countries in the world and yet is increasing economic growth at 1-9% per annum and ceasing nuclear power generation.

2. Precise figures are available at this stage for Australia based on data from the Australian Bureau of Statistics, 'Capital Stocks 1988-1989', Canberra, Australian Government Printer 1990. Work is proceeding to produce an equivalent New Zealand figure although the expectation is that the percentage will be comparable but slightly lower.

3. In the residential building sector, energy savings of 50% are relatively easily obtained by using such well-tried measures as passive solar heating and more efficient use of energy for lighting, cooking and water heating. In the commercial building sector energy savings of 20 – 25% are available by good management techniques including the use of more efficient plant and equipment. Savings of up to 50% are available by modifying the form and orientation of the building to allow for daylighting and perhaps natural ventilation.

4. Several recent local studies suggest that the capital energy of a building is between 3 and 15 times the annual operating energy of the building, depending on size, type of construction and use.

5. "Climate Change, The Consensus and the Debate", Ministry for the Environment, Wellington, 1991, suggests total emissions for carbon dioxide and methane nationally by industry sectors. However it is difficult to isolate the built environment component as it tends to influence all sectors. The tendency is to underestimate the contribution.

6. See Peter Newman, "Social organisation for ecological sustainability – towards a more sustainable settlement pattern" in *Fundamental Questions Paper No 11*, Centre for Resource and Environmental Studies, Canberra, 1990.

7. Genuine alternatives to fossil fuels for motive power are still not readily available.

8. "How long will this Green fad last?" and "If energy efficiency is so good the market will do it anyway" are often-heard arguments in New Zealand.

9. Successive local and overseas studies of occupant preferences indicate a clear desire for environment-friendly, user-friendly spaces, i.e. natural non-toxic environments.

Graeme Robertson is a Senior Lecturer in the Department of Architecture, University of Auckland. He is a member of the International Union of Architects Project Group "The Implications of the Greenhouse Effect for Architecture and the Built Environment" and national coordinator of the New Zealand Institute of Architects Environmental Policy.

Questions and Concerns

READING 2

1. If Canada's world-renowned R-2000 building standard is so well suited to our cold climate and the owners of R-2000 homes save thousands of dollars in energy costs over the years, why are so few R-2000 homes built?

2. Making a home virtually air tight is the best way to control its consumption of energy. But controlling the interior "climate" with a mechanical "air exchanger" also poses some risks. What are they?

HOUSING

R-2000

Canada has developed state-of-the-art, energy efficient house building technology. So why aren't we buying it?
BY TIM LOUGHEED

The oil crises of the 1970s bequeathed a valuable legacy to Canadian home buyers of the 1990s — some of the most energy efficient housing to be found anywhere in the world. Inspired by the prospect of lower heating costs, the federal government initiated a research program to set new standards in residential heating, insulation and ventilation technology. The result was a purely Canadian design package called R-2000, yielding a house which consumed only half as much energy as a traditional house of the same size.

Today, an R-2000 home is more likely to be described as environmentally friendly or healthier for its inhabitants. Regardless of this distinction, only a tiny fraction of Canada's housing has been built to these exacting standards. Yet the assured quality of these houses continues to make an impression on builders and buyers across the country.

The federal Department of Energy, Mines and Resources (EMR) estimates it has spent $50 million on R-2000 since the late 1970s, when work began under the Super Energy-Efficient Home program. Following the formal commercialization of the R-2000 system in 1984, with a variety of demonstration homes already constructed, the government teamed up with the Canadian Home Builders' Association (CHBA) to deliver R-2000 to its 12,000 member companies across the country.

Other related agencies have also helped cultivate R-2000 technology: the Heating, Refrigerating and Air Conditioning Institute of Canada, the Canadian Standards Association, the Canadian General Standards Board, Canada Mortgage and Housing Corporation, the Canadian Electrical Association and the Canadian Gas Association.

R-2000 — WHAT IT MEANS

Although "R" value generally refers to the heat insulation capacity of a building's surface, such as an "R-24 wall", R-2000 specifies much more. In fact, the "2000" is not a specific measure but signifies the futuristic building standard the government wanted to see in place by the year 2000. High value insulation is only one component among several enabling an R-2000 structure to meet such standards.

The technical centrepiece of the R-2000 approach is the heat recovery ventilator (HRV), a device which continually renews the indoor air supply. A typical HRV would be about the size of a filing cabinet and cost $2,500 to $3,000. Located near the furnace in the basement, this unit draws outside air into the house and expels air which has already been inside. The incoming and outgoing air streams pass by each other, evening out the temperature before the outside air enters the furnace or the air conditioner and proceeds throughout the house.

According to R-2000 technical requirements, an HRV must completely exchange the air in the house more than once an hour. Ventilation on this scale eliminates any need to open windows, since fresh air flows continuously even in the middle of winter.

This controlled ventilation enhances the effect of another key R-2000 feature: an interior framework which presents a single complete barrier against

R-2000 INCENTIVES

Practical incentives to encourage buyers of R-2000 housing are still few and far between, but they do exist.

New Brunswick Power Corporation now indirectly offers $1,000 to anyone purchasing an R-2000 home in the province. The deal, administered through the provincial home builders' association, also provides a comparable rebate for the builder.

The New Brunswick Housing Corporation has been the leading taker, specifying that all of the province's public housing must be built to R-2000 standards. That policy has created a boom among New Brunswick's qualified R-2000 builders, and a push by others to become qualified as soon as possible.

According to George Dashner, NB Power's technical co-ordinator for energy conservation, the utility is engaging in Demand Side Management (DSM). He says the amount of the rebates is based on long-term savings the company anticipates by taking on new R-2000 customers rather than serving conventional homes.

Dashner notes that almost 10 per cent of housing starts in New Brunswick are R-2000. He credits this relatively high proportion with the small size of the province's building community, which has enabled supporters of R-2000 to reach almost everyone who could be involved.

Ontario Hydro is also practising DSM, offering $2,000 to R-2000 buyers in areas not served by natural gas heating; another $2,000 goes to the builder. Again administered through the province's home builders' association, the program is set to remain in place until the end of 1992, when it will likely be extended.

The utility also offers an incentive to Ontario home owners who want to upgrade their windows, compensating them for the added cost of improvements such as high efficiency gas-filled panes.

Other provincial electric utilities, in British Columbia, Manitoba, Prince Edward Island and Nova Scotia do not offer incentives but have committed to working with EMR and CHBA to promote R-2000.

Among financial institutions, the Bank of Montreal is working with provincial home builders' associations to reduce mortgage rates to R-2000 home buyers. The reduction is not a direct subsidy but depends on terms established with the builder, which can drop the interest rate by as much as one quarter of a point.

The bank's program is now effective in New Brunswick, Nova Scotia and British Columbia, with plans under way to establish a similar arrangement with home builders in Ontario. As financial institutions, builders, utilities and other interested groups join forces to bring R-2000 into the marketplace, such incentives will probably pave the way.

the leakage of air or other vapours. By lining walls and ceilings with a single sheet of polyethylene or tightly joined drywall, virtually all leakage is stopped. Besides dramatically increasing the insulation value of these surfaces, the barrier prevents interior humidity from penetrating into the house frame, reduces the amount of unfiltered dust or pollen entering from outside and generally cuts down on noise.

R-2000 builders even take into account the physical site of the home. Squared along a north-south axis, windows and solar collectors are preferred on the south side — receiving the most sun — and discouraged on the north side. Roof overhangs are measured to shade windows in the summer, when the sun is high in the sky, but still allow sunlight in during the winter months. Even the possible insulation boost of a slope or a hillside can contribute to R-2000 performance.

No R-2000 "look"

In spite of such intricate details, however, R-2000 program manager Jack Cole stresses that such houses remain largely indistinguishable from traditional models.

"It's not meant to look any different," he says. "It doesn't have to be built into a berm or somewhere up in the woods."

In reality, practically any house plan can accommodate an R-2000 approach. A computer program developed by EMR helps builders do just that, providing a simulation of the heating and hot water costs of a particular design. Besides determining whether the proposed house will meet R-2000 standards, this computer modelling procedure enables designers to fine-tune their plans with a quick assessment of how alterations might affect the building's energy consumption.

Builders need special registration to do any R-2000 work, which requires them to take a course updating their knowledge and outlining the technology. Each builder must then complete a trial house, undergoing a thorough inspection by R-2000 inspectors during and after construction. Some 4,000 Canadian contractors have become qualified to build R-2000 homes, although only around 400 may be doing so at any given time.

In fact, for all their demonstrable superiority and cost effectiveness, Canada still has only about 6,000 R-2000 homes. By comparison, about 150,000 to 200,000 new homes go up across the country every year. That discrepancy might be cost related, since an R-2000 home can cost more to build than a comparable conventional house. But the difference is often no more than one or two per cent of the price of the house, and sometimes there is no difference at all.

Many builders, too, have likely been unwilling to take the trouble to abide by the administrative process imposed by R-2000 construction. Some of these same contractors might well have applied R-2000 techniques and erected houses that would qualify or come close to qualifying as R-2000, but without an inspection and certification no official record exists. There may be three or four times as many of these R-2000 "clones" as there are certified houses.

R-2000 INFLUENCES BUILDING TECHNOLOGY

According to Steve Carpenter, president of the Kitchener-based consulting firm Enermodal Engineering, the technically demanding nature of R-2000 housing falls squarely on the builder.

"It's just not consistent with building whole subdivisions of houses," he says. "If you want an R-2000 house, you're favouring a small custom builder over a large tract builder."

R-2000 houses must be built from scratch. Although significant improvements can be made by adding R-2000 features to an existing home, getting the structure to certifiable standards would involve almost complete rebuilding. Although the R-2000 program may eventually expand to include this kind of "retrofit", the available market remains limited to new housing.

Nevertheless, the Canadian Home Builders' Association suggests that R-2000 has made its presence felt across the industry. Robert Sloat, CHBA's technical director, says the program was not necessarily intended to produce large numbers of high performance dwellings. Instead, he portrays R-2000

R-2000 IN JAPAN AND ELSEWHERE

It came as no surprise to R-2000 builder Ray Pafford that the Japanese would be interested in adopting Canada's R-2000 housing technology. Wearing an R-2000 lapel pin makes him a constant target of earnest inquiry from builders everywhere he goes.

"When I go to conferences down in the States, at coffee break time I just don't get a chance to get a hot cup of coffee," he says, describing the seemingly insatiable curiosity builders around the world have with Canada's high quality housing. "I don't mind. I love it."

Japan's 2 X 4 Home Builders' Association acted on its curiosity in 1990, when it entered into a licensing agreement with EMR to use R-2000 technology, including the trademark and logo. The Japanese plan to establish their own R-2000 program, which will have to meet material, equipment and performance standards set in Canada.

In a similar move, the Alaska state government and the state's home builders' association have developed a housing program which duplicates almost every aspect of the R-2000 technology, training and implementation. Although not a direct appropriation of R-2000, Alaska's Craftsman Home Program did take advantage of expertise provided by Canadian R-2000 builders.

Robert Sloat, technical director of the Canadian Home Builders' Association, accepts such imitation as a very sincere form of flattery.

"We have demonstrated a world-beating technology," he says. "We don't have to take a back seat to anybody when it comes to housing technology."

as an effective way to set higher standards and introduce new construction technology.

"It has fostered the development of new and innovative approaches which have become mainstream," he says. "The issue is not to create a standard low enough to create a large market share. Rather it's just the opposite — to continue to maintain the leading edge so you'll always have this technology transfer function."

Cole suggests virtually all new housing incorporates some aspects of R-2000 design, enhancing the quality of a wide range of homes.

"R-2000 is a voluntary standard," he says. "It's meant to draw the building code along. We call it a leading edge standard and it's always meant to be ahead of the pack. If every house in Canada was R-2000, we'd have to say it's time to move R-2000 up one more notch. We don't want to be in the mainstream."

Cole confesses that attention to the research and development of R-2000 has come at the expense of marketing. "People have criticised us in the last few years, saying we're stressing the technical side too much, we're not promoting it well enough. The consumer has been lost in the technical side."

Lost or not, Canadian consumers have relatively privileged access to some of the most advanced housing innovations available anywhere. Prospective buyers should at least consider R-2000 as an issue, if only because it provides a useful vantage point from which to examine other significant housing issues: cost, health and environmental impact.

CALCULATING COSTS

Since a house represents the single largest purchase most people make, cost looms largest in importance. Calculations should not stop with the price of the home, but extend to maintenance and operating expenses. In the case of a used home, buyers can gauge these expenses by asking for copies of utility bills covering a period of a year. New homes present more of a problem, and buyers could end up guessing about their month-to-month expenditures based only on a builder's estimate.

Because an R-2000 house's performance was simulated by computer during the design stage, these expenses can be reliably forecast. Indeed, they must fall within certain limits for the house to qualify as R-2000.

If a house is being touted as an R-2000 house, the trademarked R-2000 logo should be evident somewhere, such as on the main electrical panel door. There should also be a copy of the R-2000 homeowner's manual, which explains the system in detail. If these items are missing, a buyer should question whether the house was inspected and approved. If not, then the house — which might be very well built and perfectly adequate — can offer none of the assurances of R-2000 technology. Those assurances are considerable, but they accompany the program exclusively.

GROWING PAINS

Like any innovation, early R-2000 installations did encounter problems. For example, improper mixing of warm and cold air streams led to frost developing in some HRVs. Similarly, the first generation of HRV fans tended to be noisy. Such problems were solved early on,

R-2000 HOME OWNERS

Ray Pafford was one of the earliest R-2000 home owners — and he remains one of the most enthusiastic.

"It's quiet, it's clean, it's bright," he says. "It's an even temperature throughout. The energy costs are good. It's just a nice house."

Pafford, a builder himself, put up his R-2000 home in Corner Brook, Newfoundland in 1983. Since then he's been an active R-2000 builder and inspector, as well as introducing the system to other builders.

With lots of windows, a solarium and solar collectors on the south side — and one quadruple-glazed window on the north — he enjoys the house's low energy costs.

"I'm just one of those guys who doesn't like wasting anything," says Pafford, noting his annual electric heating bill has yet to top $450.

But he adds what really drew him to R-2000 was the high quality of design and greater attention to detail. He shares his disappointment that R-2000 homes have not become more popular. He concludes the program got off to a bad start when many of the first R-2000 homes turned out to be large and expensive, suggesting to some observers that this innovation would be costly.

Yet for Janet and Jim MacDonald, of Trenton, Ontario, R-2000 added no extra cost at all. Five years ago when they considered the design of a conventional bungalow with a two-car garage, R-2000 added nothing to the final price. They have been pleased with their choice.

"We love it," says Janet, giving the example of fresh indoor air. "It's good clean air all the time. We never open windows."

Opening windows is usually not an option for Erick Oles in Whitehorse, Yukon, but the air in his R-2000 home remains just as fresh. When he built his 4,000-square-foot home five years ago, it was one of the largest R-2000 dwellings in the territory. Comparing heating costs with a neighbour whose house is just as big, Oles notes his bills come out to about half as much.

As a builder, he especially admires EMR's computerized design program, which enabled him to "fine tune" his house for maximum efficiency. He admits he found the actual construction standards exceedingly strict, but he still praises the impact R-2000 has had on the housing industry.

"The essence of the R-2000 program was to open the eyes of builders who were willing to look at new technology and build better homes from that point on," he says. "I certainly did, and I think the vast majority of us have done that."

however, and few others have appeared.

"They didn't have all the bugs worked out," says Enermodal's Carpenter. "Five years ago, the whole R-2000 concept was a little foreign, some of the products were new and people didn't know what they were doing. It's matured enough now — a lot of that has disappeared."

In this light, buyers should beware of "almost R-2000" houses presented as cheaper than a similar certified model. Eliminating a component such as the HRV will certainly lower the price — as well as ongoing operating expenses — but if a secure vapour barrier is in place, the results can be disastrous. Lacking proper air circulation, such houses will retain their humidity, presenting health problems such as the spread of moulds.

Moreover, an incorrectly installed HRV can be out of balance — drawing more air from the house than it brings in. This negative pressure could be significant enough to draw toxic fumes from a fireplace or furnace into a room's atmosphere. Because of such dangers, HRV installations should be certified by the Heating, Refrigerating and Air Conditioning Institute of Canada. This certification is automatic with an R-2000 home, but can also ensure that any installed unit meets similar standards.

Recognizing Long-Term Benefits

Regardless of the type of house, the initial cost associated with any aspect of R-2000 technology must always be weighed off against long term benefits. Once again, such comparisons are easier with a certified R-2000 home, where energy use can be reliably predicted.

"What I don't like is a builder selling the technology on it being an economical thing, when it may be priced at such a level that it is no longer that," says Bruce Sibbitt, an energy consultant with Mississauga-based Caneta Research. "I would look carefully at what you're being asked to pay for these benefits, and compare that with the real estate and what your savings will be."

Prospective buyers should also expect similar comparisons to affect the question of resale value. Although realtors encounter R-2000 too rarely to make definitive judgements, there is every reason for a certified home to possess an edge over conventional housing. In the Maritimes — where the predominance of electric home heating has created the country's most dynamic R-2000 market — surveys point to resale prices as much as five per cent higher than similar conventional homes.

Even banks may be persuaded to grant a higher value to an R-2000 house. The Bank of Montreal has expanded its "PIT" formula for calculating mortgages — incorporating the monthly Principal, Interest and Taxes on the house — into a "PITE" formula including monthly energy costs.

In the case of an R-2000 home, this added factor could enable prospective buyers to obtain a higher mortgage, based on significantly lower heating expenses. Working in conjunction with provincial home building associations, the Bank of Montreal will drop mortgage rates up to a quarter point to R-2000 buyers in three provinces.

Certain non-economic considerations also bear on the purchase of a house. In the case of an R-2000 house, indoor air quality is an attractive bonus. The HRV, by exchanging air often and thoroughly, removes obvious airborne contaminants like cooking odours, dust or pollen. More subtly, this system also inhibits the growth of moulds and reduces the levels of gases such as formaldehyde, which emanate from glues used in plywood.

In a still larger context, R-2000 proponents now refer to important environmental gains. With less energy going into heating — be it gas, oil, wood or electricity — the result is less pollution. Curiously, research dedicated to energy conservation — spawned by a series of rude oil price shocks — fits in nicely with today's general policies focusing on environmental conservation.

Such flexibility suits R-2000 proponents just fine. They are already looking past the current standard to even more sophisticated house designs which will consider more energy efficient appliances, minimizing waste water and recycling of materials. For now, R-2000 represents Canada's answer to advanced housing.

"We want to be able to say that when you buy R-2000, you're buying what is a certified, quality assured product," says EMR's Cole. "You're not just going to get what may be a good house — you can count on it."

Tim Lougheed is an Ottawa-based freelance writer.

Advanced Houses Program

While R-2000 houses continue to leave their mark on traditional home construction, builders and designers are already looking ahead to the next generation of improved housing technology. Co-ordinated by the Canadian Home Builders' Association and the Canadian Centre for Mineral and Energy Technology, a research and development branch of Energy, Mines and Resources Canada, the Advanced Houses Program has attracted millions of dollars in sponsorship from industry, power utilities and provincial government departments.

Participants from across Canada formed teams to draft proposals for houses employing the latest improvements in energy consumption and environmental impact. Each design had to project only one-quarter the energy consumption and one half the water consumption of a similar conventional house. Adopting features such as wind and solar power assistance, highly efficient appliances, and an emphasis on renewable resources, these advanced houses are intended to exceed the high standards already set by R-2000 technology.

The Advanced Houses Program, says Robert Sloat, CHBA technical director, serves as a functional laboratory where builders, designers, manufacturers, suppliers and other interested parties can experiment with new ideas to further the drive toward better housing.

Questions and Concerns

READING 3

1. The author of this article has managed to achieve energy self-sufficiency at his farm. He requires no electricity from the electrical utility. Do you think his approach is realistic for others? What are the most attractive and least attractive components of his home energy system? Why does the environment benefit? Isn't electricity a "clean" source of power? If not, why? What is a clean energy source?

Home power

Towards an environmentally friendly lifestyle

Milton Wallace

Like many other readers of *Alternatives*, my wife and I have worked for years on environmental issues. Two years ago we left Toronto and moved into a 100+ year-old log house with no plumbing or electricity on 20 hectares of fields and woods. This move suddenly brought us face-to-face with planetary environmental problems in miniature. Could we — or anyone — turn significant portions of the environmental principles we had espoused into practical reality?

One of our most immediate needs was electricity to run our computer to make our living. It didn't seem right to connect our 19th century home to a nuclear reactor, so we decided to build our own 21st century power system.

Renewable energy power production is much easier on the environment than is connecting up to Hydro. It doesn't contribute to the serious problems of global warming and acid rain. Dr. Raye Thomas, president of the Solar Energy Society of Canada, in a submission to the federal Standing Committee on Environment, talked about how commercially available active and passive solar, and photovoltaic technologies can reduce conventional fossil fuel emissions. Dr. Thomas discounted the nuclear option, saying, "We believe that nuclear energy is not really a viable alternative [to burning fossil fuels], because the processing of uranium fuel produces carbon dioxide. We do not believe it is cost effective. It has low public acceptance, serious waste disposal problems and the potential for radiation leaks."

Bringing hydro power into our house would have required that we clear-cut an eight-metre wide corridor through our woodlot — a big impact on the local ecology, visual pollution forever, and a few less trees to help keep the global temperature down.

The first step in implementing a commitment to an alternative power system is to decide if you can afford it. To minimize the cost, it's wise to examine all your energy usage. Drawing power from the power company can lead you to generous electrical usage, including heating some or all of your home electrically. Most people who produce their own power tend to use electricity conservatively and only when some other form of energy won't work. Table 1 shows the approximate usage of different forms of energy at Sun Run Farm.

Any human activity has some impact on the environment and alternative energy production is no exception. Actually the main environmental impacts of solar panels and other system components occur during the production process, from conventional fuel combustion and process emissions. Once installed the alternative system is virtually pollution-free.

Courtesy of Milton Wallace

Table 1
Energy system by household task

Electricity:
- solar   60 percent
- wind   25 percent
- propane   15 percent

Heat:
- wood   95 percent
- propane   5 percent
- retrofitted thermal windows
- R40 ceiling, R20 walls
- entensive work to prevent infiltration
- windbreak plantings
- proposed greenhouse
- passive solar addition

Hot water:
- propane   100 percent
- R20 insulation on water tank
- all hot water pipes insulated
- proposed solar hot water heater
- anti-convection loop

Cooking:
- propane   50 percent
- wood   50 percent

Refrigeration:
- propane   90 percent
- electric   10 percent
- 5 cm insulation on all sides of refrigerator
- ammonia working fluid

Household cooling:
- no air conditioners
- mostly natural cooling – deciduous trees on south side,
- cool air flow from cellar

Clothes dryer:
- solar (outdoor lines)
- wood heat (indoor lines)

Transportation:
- fuel efficiency: mostly at 45-50 mpg (Pontiac Firefly)
- reduction: work at home – less commuting
- future goal: solar/alternative fuel hybrid car

Food:
- mostly organically grown at home – compost all organic waste

Waste:
- recycle newspaper, fine paper
- cardboard, cans, glass, mixed plastics, copper, aluminum, oil

Fossil fuels used in the back-up generator are, of course, polluters. We chose propane because it is multi-purpose and burns cleaner than gasoline or diesel. It is reported that hydrocarbon emissions from propane usage are 50 percent lower than those from gasoline and have lower ozone-forming potential. Carbon monoxide production is considerably less. We hope that our future building plans and system optimization will greatly reduce propane usage.

Wood for fuel comes from our farm and is harvested in a sustainable manner. Although carbon dioxide is released when wood burns, it is also absorbed from the atmosphere as new trees grow to replace those harvested. The main wood stove is very efficient and meets EPA standards for emissions.

Independent power systems can be tailored to meet your specific requirements. We selected a mix of generators including sixteen 50-watt photovoltaic panels, a 250-watt wind generator on a 15-metre tower, and a 6.5-kw propane generator for backup.

Solar power production is good through much of the year. Two solar pluses in Ontario are greater panel efficiency at cooler temperatures and enhanced power production due to snow reflectivity. However, solar production drops off significantly during the overcast days in November, December and January. The wind generator helps make up this deficiency because wind power production is greatest in the winter due to higher wind speeds and lower wind resistance from leaves and other vegetation.

Power from all three sources charges a large battery bank consisting of six lead-antimony cells, each weighing 145 kg. Fully charged, they will run the house and office for three to five days. The battery cells are guaranteed for 20 years and may last longer than that, in our system. A 2000-watt inverter converts 12-volt DC battery power to 110 AC, the standard house current.

Our power system is also owner-financed. We built the system gradually over a couple of years and so were able to spread out our costs. Total system cost was comparable to that of a new car. It is interesting to note that the cost of Ontario Hydro's 25-year demand/supply plan is enough to build an alternative power system for every family in Ontario.

As we are finding out at Sun Run Farm, there is a safe and environmentally friendly alternative to expensive public power plans based on risky and environmentally unsound technology. It is possible to build a reliable electrical power generating system that meets all your needs. Knowledgeable sources are saying that the 90s are going to be for photovoltaics what the 70s and 80s were for electronics and computers. The future for home power has never looked better.

Milton Wallace is an environmental consultant, and he designs and distributes alternative energy systems under the business name of Sun Run Energy.

Questions and Concerns

READINGS 4 THROUGH 7

Readings 4 to 7 have been excerpted from a manual published by Canada Mortgage and Housing Corporation (CMHC, a federal government corporation) for architects and designers who entered CMHC's Healthy Home competition. Each section gives a brief summary of an issue relating to housing and the environment.

1. The section on indoor air quality mentions contaminants "built" into buildings. These include the commonly used bonding agent formaldehyde, which is found in carpeting, chipboard (used to make cabinets and furniture), and hundreds of other products. A known carcinogen, formaldehyde can "outgas" from the products in which it is contained. Think about all the materials in the room around you (and other rooms you spend time in) and try to identify those that might give off unhealthy gases and those that are inert.

2. We take the quality of our drinking water for granted in most parts of Canada. Municipal tap water is regulated by the Canadian Water Quality Guidelines issued by Environment Canada. But these guidelines allow for "acceptable levels" of many contaminants. How would you define an "acceptable level" of a cancer-causing chemical? Do you think scientists can be sure about acceptable levels of dangerous substances?

3. Electro-magnetic radiation is emitted from power lines and from many devices in our homes and schools. List the places where you might encounter electro-magnetic radiation during a typical day. Should people be more concerned about it?

CMHC • HEALTHY • HOUSING • DESIGN • COMPETITION • GUIDE • AND • TECHNICAL • REQUIREMENTS

The Issues
1.0 Occupant Health

Concerns relating to the impact of housing on occupants are being increasingly voiced across the country. Disorders ranging from asthma and allergies to immune system disfunctions and chemical hypersensitivity are being linked to the quality of the indoor air. To date, the healthfulness of housing has been addressed on an issue by issue basis and from the perspective of isolated disciplines. Medical practitioners attempt to draw linkages with environmental factors, environmental scientists probe the relationships between the built environmental and health, air quality specialists attempt to measure and identify contaminants in the indoor air, and mechanical engineers attempt to design ventilation and air treatment strategies.

Health, itself, is complex and holistic — dependent on hereditary, dietary, emotional, psychological and environmental factors. Consideirng only the environmental factors, it is evident that health is affected by airborne pollutants and toxins, moulds and mildews, particulates, humidity levels, ions, radioactive elements, light, electromagnetic fields, thermal conditions and sound — to name but a few. An improved understanding of the contaminants and pollutants causing these disorders allows designers and builders to now specify healthier housing — housing which accommodates the desire of homeowners for safer, healthier homes.

In designing a 'Healthy House,' a holistic approach to occupant health must be pursued — balancing all factors. The design team must consider indoor air quality, water quality, and background factors including light, noise and electromagnetic radiation.

1.1 Indoor Air Quality

As envelope tightness is increased, natural infiltration rates decrease, leading to the potential for greater concentrations in the home of contaminants and pollutants and reduced fresh air for the occupants if remedial strategies are not put into place. A Healthy House will ensure the occupants of good air quality and an adequate supply of fresh air.

While a Healthy House may not need to incorporate all of the strategies required to meet the needs of the hypersensitive, many of the strategies may be employed for the general population. Several guiding principles should be considered when designing healthy housing:

- Reduction of the level of contaminants 'built' into the building;
- Removal of any contaminants at the source of production; and
- Dilution of house air with fresh outside air.

Reduction

The amount of potential contaminants incorporated into standard building materials is significant. Common contaminants causing adverse occupant reactions include:

— volatile organic compounds (from manufactured wood products, carpets, paints, household cleaners, fabrics, inks etc.)
— petroleum (oil and gas vapours)
— moulds, dusts, pollens, animal dander
— woods (natural resins from pine, cedar etc.)

The most frequently reported reactions to these contaminants include tension fatigue, headaches, and eye, ears, nose and throat irritation. In some instances the symptoms are severe enough to interfere with a person's daily activities, life and career.

Materials, construction systems, and mechanical systems should all be evaluated based on characteristics including outgassing, stability under

Reproduced from: Government of Canada, 1991, *The State of Canada's Environment.*

exposure to varied temperatures and moisture levels, cleanliness, maintenance requirements and durability.

Moisture, and its relationship to mould and mildew generation has also been recognized as a contaminant with serious health implications. Construction techniques and mechanical systems which ensure acceptable interior comfort levels, while minimizing health related problems, are a cornerstone of a Healthy House.

Removal

Any combustion by-products resulting from the operation of fossil fuel or wood burning appliances must be vented directly to the exterior, without risk of spillage into the interior environment. Similarly buildings in areas with high radon levels should be designed to remove these gasses before they enter the house.

In addition, homeowners are responsible for a significant component of the production of contaminants, pollutants and irritants in the home, including: moisture production, odour production, and the use of cleaning products, as well as contaminants produced in hobby related activities. A Healthy House will allow the homeowner to remove any contaminants at the source of production.

Dilution

Bringing in fresh exterior air is the third stage in a healthy indoor air quality strategy. In considering ventilation strategies which exchange fresh air for exhaust air, the design team should ensure that incoming air is brought into the home in as clean a manner as possible, in the quantities required for a healthy interior and that the fresh air is thoroughly distributed to all areas of the house.

Strategies to address indoor air quality must be balanced with the need to ensure acceptable comfort levels for the house occupants. The distribution of fresh air through the house must be designed to ensure humidity and temperature control within acceptable comfort limits.

Summary

Many of the features in "low-pollution" housing are designed to meet the needs of the chemically hypersensitive and individuals with respiratory problems and allergies. They may also be applicable to housing for the general population. Reduction and removal strategies to consider, include:

— heating systems with minimal spillage of combustion byproducts; low temperature heating systems are preferable;

— hard-finish flooring such as ceramic tiles or hardwood; tiles can be laid with cement mortars rather than adhesives; concrete without admixtures, water reduction oils and curing agents, can be used for foundations;

— building materials with no formaldehyde or minimum emission of volatile organic compounds; woods should not be treated with preservatives;

— wall and ceiling finishes that do not require paints (such as plaster), or if painted, nontoxic paints are used;

— draft free building techniques designed to reduce the infiltration of contaminants from the outdoors or from materials in the building envelope;

— good outdoor ambient air quality and location away from heavy traffic, industrial pollution, or power lines is emphasized;

— a ventilation system to bring in fresh air and exhaust stale air from local sources of pollution within the house;

— an air purification system to remove airborne contaminants such as dusts, mould spores, pollens and chemical pollutants;

— a central vacuum system which exhausts to the outside, or other suitable means of removing dust from the home;

— furniture, furnishings, household products selected for minimum emission of volatile chemical contaminants;

— a sufficient amount of natural lighting; and

— a high degree of care during construction to minimize dust and other contaminants.

CMHC • HEALTHY • HOUSING • DESIGN • COMPETITION • GUIDE • AND • TECHNICAL • REQUIREMENTS

1.2 Water Quality

Until very recently the availability of pure drinking water was taken for granted in most parts of Canada. Growing awareness of industrial pollution, the limitations of municipal water treatment and outdated infrastructure have called into question the source, the treatment methods and the distribution system presently in use.

Potable water is obtained from surface water (rivers and lakes) or from ground water (wells and springs). Whatever the source, water may be contaminated by bacteria, by chemicals or by metals. Municipal water treatment systems were developed in response to the awareness that diseases such as cholera are spread through contaminated water.

Recent concerns about the quality of drinking water derive more from awareness of industrial and agricultural pollution of surface and ground water supplies. As well, there are some concerns about the health effects of chemicals used in the treatment process, (chlorine in particular) and of the leaching of heavy metals such as lead from the distribution system.

These concerns are reflected in the growing use of bottled water and point-of-use home treatment systems. However, there is a lack of applicable standards to guarantee the quality of either bottled water or the effectiveness of home treatment devices. In comparison, municipal tap water is regulated under the Canadian Water Quality Guidelines, Environment Canada, 1987, and is still the best choice in terms of overall water quality.

Where municipal water supply is not available, treatment methods should be carefully reviewed for effectiveness, servicing requirements and safety.

Home Treatment

Home treatment for removal of bacteria should only be required where the water source is independent of municipally treated supply. Technologies for disinfecting water in the home include:

- chlorination,
- iodination,
- distillation,
- filtration with ceramic filters,
- ultra-violet irradiation, and
- ozonation.

Iodination and chlorination are the only methods which can provide protection against the build-up of micro-organisms in the distribution system. All of the methods require careful attention to operation and maintenance for their safe use.

Other methods are available to remove chemicals and metals. Activated carbon filters have been shown to be effective in removal of trace chemicals. They are sometimes used as part of a two-stage home treatment system to remove chlorine, iodine or ozone residuals introduced during the disinfectant process. Reverse osmosis devices have been shown to be more effective in the removal of metals.

Both treatment devices require periodic changing of a filter or membrane to prevent bacteriological contamination or "breakthrough" of contaminants. Some home treatment devices will greatly increase water consumption.

Reproduced from: Government of Canada, 1991, *The State of Canada's Environment*.

CMHC • HEALTHY • HOUSING • DESIGN • COMPETITION • GUIDE • AND • TECHNICAL • REQUIREMENTS

1.3 Light, Sound and Radiation

Health scientists and consumers are beginning to voice concerns over the effects of light, noise and electro-magnetic fields on human health and well-being.

Light

The relationship between sunlight, vitamin D and bone growth has been known for generations; children in northern countries are customarily given vitamin D supplements during the winter months. More recently, doctors treating patients for Seasonal Affective Disorder (SAD) have found that symptoms are alleviated by exposing the patient to more sunlight. Specialists involved in workplace design have also found that performance is enhanced if workers have access to daylighting.

Fortunately, the aim of introducing more daylight into the home coincides with energy efficiency. And, the advent of high-performance windows means that designers can plan for generous daylighting — even in rooms with northern exposure — with a lower penalty in thermal performance. The most effective daylighting strategies use windows on two sides of a room to reduce glare. For larger buildings, new technologies such as light pipes or solar assisted light wells are available to bring daylight to the interior of the building. Full spectrum artificial lighting will enhance interior environments.

Noise

Noise has been referred to as the next pollution issue. Many noises in the urban environment are sufficiently loud to cause hearing damage, not just annoyance.

Building technology is available to reduce noise experienced in the home from both internal and external sources. Thicker envelope construction as for R-2000 homes has the side benefit of reducing noise transmission — isolating the interior from outside noises. Improved detailing and construction practices can reduce transmission within the house and between semi-detached and row-house units.

Air borne sounds can be minimized by sealing any air leakage paths between rooms. Sealing around electrical outlets, plumbing, and penetrations through and under walls will reduce airborne sound transmission.

Sound absorbing materials can be used to isolate a particular noise source such as a hobby room or TV room. Fibrous materials (mineral, glass or cellulose), work effectively to reduce sound inside an enclosure, a wall, or a room. Sound barrier materials (commonly drywall, plywood, concrete or glass), are non-porous and solid, reducing sound energy passage through reflectance. The heavier and thicker the material, the greater the sound reduction.

Plumbing systems, house appliances and fans (exhaust, furnace, heat recovery ventilators, etc.) can all contribute to indoor noise levels. Equipment which vibrates should not be directly affixed to the structure — acoustical isolators can reduce vibration and motor noise. Thoughtful design of floor plans — considering the location of equipment and appliances — and specification of quieter equipment can provide major benefits to the house occupants.

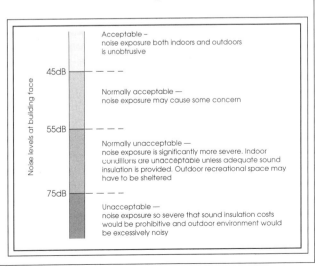

Reproduced from: Government of Canada, 1991, *The State of Canada's Environment.*

CMHC • HEALTHY • HOUSING • DESIGN • COMPETITION • GUIDE • AND • TECHNICAL • REQUIREMENTS

2.1 Embodied Energy

Manufacturing Energy Inputs

Recent research has focussed on the energy embodied in the materials used in the house construction process. It has been estimated that the energy embodied in the materials — the amount of energy required to manufacture, transport and install materials used in house construction — represents as much as 30 years worth of operating energy consumption. As this research is refined further to account for local and manufacturer specific processes, designers and builders will be able to make choices relating to house design and material selection which optimize energy use.

A considerable amount of energy is consumed in the fabrication of materials used in a typical house. Large quantities of energy are required to produce everything from glass for windows and glass fibres for insulation through to bricks and ceramics. Major energy savings can be made through the wise choice of building materials. For example, a typical wood frame home has 1/3 the embodied energy of the same house built of steel and concrete.

Material	Energy Intensity*
Concrete (ready-mix, regular weight, 2000 psi)	2742 MJ/m^3 (2096 MJ/yd^3)
Framing lumber (2 x 4 wall studs)	3264 MJ/m^3 (28 MJ per 2 x 4 stud)
Gypsum board (4 x 8 12 mm sheet)	60 MJ/m^2 (176 MJ per sheet)
* Values shown are based on averages for the Canadian economy using 1984 statistics, and include all the energy required for extraction of raw materials, processing, fabrication, and transportation to site in an urban centre. Values shown do not include installation on-site, repair and replacement over the lifetime of the home, or demolition and disposal.	
Table: Embodied Energy	

Source: CMHC and Sheltair Scientific

The accompanying chart lists typical values for energy consumed in the fabrication of construction materials.

When considering the design of a Healthy House, the design team must evaluate the energy consumed in manufacturing a product in relation to its life expectancy, to the energy saved in house operation and to the replenishment cycle of the material. Energy intensive materials with long life expectancies may be justified. While many of the insulation materials are energy intensive in their manufacturing, they can be long lasting while dramatically reducing the operating energy needs of the home. As an example, 5.1 GJ more energy would be embodied in a 2x6 (RSI 3.5) wall than in a 2x4 (RSI 2.1) wall. However the energy savings resulting from the upgrade would result in an energy payback of only 2.6 years — and significant net savings over the duration of the building.

Reproduced from: Government of Canada, 1991, *The State of Canada's Environment.*

Discussion for The Built Environment

1. Consider the concept of "embodied energy" – that all materials contain energy used in their creation or manufacture. As inert as a cement block may appear, great amounts of heat were needed to form the cement. Likewise, the making of steel, aluminum, glass, and ceramic require plenty of energy. Timbers, grown over long periods of time, embody the energy of the sun.

In the case of manufactured materials, this energy also implies the release of greenhouse gases like carbon dioxide, and even more carbon dioxide when they are transported. If they have to be replaced, when a building is demolished, their embodied energy is lost and new, energy-consumptive materials must be installed in their places. Look around you and discuss the embodied energy that exists in the materials you see. If embodied energy was viewed as a valuable commodity, how would it change the way buildings are constructed? Why do we not value embodied energy? Do you think our growing environmental awareness will change the way we look at the materials around us? What about "embodied pollution?"

2. Another relatively new concept, and one related to embodied energy, is the analysis of a product or material based on all aspects of its "life cycle," from manufacture to disposal. If we look at the common plastic polyvinyl chloride (PVC), for example, it starts life as several powerful chemicals combined to form a relatively inert plastic, used to make plumbing pipes and hundreds of other everyday items. But in a fire, PVC once more becomes toxic. So the beginning and end of its life cycle are potentially hazardous. Or consider brick. The clay must be quarried by heavy machinery, then transported, then formed and fired, then transported again before being used. While not chemically hazardous, brick is "energy intensive" throughout the first stages of its life. But thereafter it's a durable, long-lasting material.

Discuss the life cycles of other building materials, or things in the room around you and try to determine their environmental benefits or drawbacks.

3. A third important new concept related to both embodied energy and life-cycle analysis is "full-cost accounting." When we account for things strictly in dollars we ignore their environmental costs. Cheaply built homes that are energy inefficient may cost less today but end up costing much more over their useful lives because of the energy they squander. This lost energy is a direct monetary cost to the homeowner and an environmental cost to the rest of us, a huge environmental cost when you consider the millions of inefficient homes there are in the world.

Think about the way homes are designed and planned in a typical suburb. Low-density suburban development has a high environmental cost. The large homes consume more energy. The big lawns need pesticides and herbicides, energy for lawn mowers, and irrigation. And people who live in suburbs must drive to work or to go shopping. Compare these "costs" to those of higher density housing in the inner city, or attached units such as townhouses and apartment blocks.

Discuss full cost accounting as it applies to the buying of a home. Is an older, less expensive home a better bargain than a new R-2000 home of the same size priced $10,000 higher? How are banks starting to account for the energy efficiency of homes on which they hold mortgages?

4. The author of "Home power," the article concerning solar power in a remote home, is clearly in favour of "unplugging from the grid" and producing his own energy, with the aid of the sun, the wind, wood from his property, and by using propane gas instead of electricity. Do you think this kind of energy independence is the way of the future? Might we someday dismantle power transmission lines and nuclear generating stations and draw our power from renewable sources like the sun and wind? Why are these sources of energy more environmentally sustainable than the burning of coal or nuclear fission? What are some of the economic benefits that would arise from energy self-sufficiency?

Introduction

Biodiversity

This issue examines the value of biodiversity among plant species, and how the continuing worldwide depletion in biodiversity threatens global food supplies and the development of valuable plant byproducts such as medicines. We also look at the work of two Canadians who are helping to defend and rebuild the Earth's dwindling plant genetic resources.

Strolling through a grocery store you would never know that there was once over 30,000 varieties of rice grown in India or that hundreds of varieties of potatoes are still found in Peru. The food we buy in North America normally comes from a relative few standard varieties of wheat, rice, potatoes, tomatos, corn, and other vegetables bred for qualities such as rapid growth and their ability to withstand long-distance transportation from field to supermarket.

Modern centralized food production – which has moved the Canadian food supply from the hands of hundreds of thousands of small farmers a century ago into the factories of several hundred large "food processors" today – has sacrificed much of our inherited species diversity. The same problem exists across the world. India now gets 75% of all its rice from a mere 10 varieties. The many thousands of other varieties have become rare or extinct. It is estimated that at present rates of extinction, about 60,000 plant species – roughly 25% of the world's entire remaining plant inventory – may be lost or endangered within the next 50 years. By some estimates, we have already allowed 75% of genetic diversity in the world's 20 key food crops to be lost forever during the last 50 years.

Why is this a problem? Each strain of any plant species evolves over time to possess a unique genetic "personality," including certain genes developed in reponse to habitat conditions. These "encoded" qualities might be a resistance to certain pests or diseases, the ability to grow under drought conditions, a richness of taste, or the ability to manufacture a rare chemical that might be extracted for use as a human medicine. (About 25% of all modern pharmaceuticals are derived from plants, including an extract from Madagascar's rosy periwinkle that is part of a drug that fights childhood leukemia, another cancer-inhibiting extract from the bark of the Pacfic yew tree, and a substance recently found in the Chinese cucumber that might work against AIDS.)

Especially valuable are the genes that arm one variety of a plant with natural immunity from diseases that afflict other varieties of the same species. When growers in a region use different varieties of the same plant, chances are disease will only harm part of the total crop. In this way, maintaining sufficient diversity is a hedge against wholesale crop damage, a way to protect our supply of food. Reduce diversity, however, and a single blight could theoretically wipe out the corn or wheat crop of an entire region if all varieties were the same or similar. This happened in 1970 when a virulent fungus killed half the corn crop from Florida to Texas. In the words of author Robert Rhoades, as he held some unusual grass seeds from a remote Siberian river valley (Reading 1), "Could one of these grains contain a gene for a minor agricultural revolution? Like the wheat... from Turkey with a disease resistance worth 50 million dollars annually to the U.S.? Or the Ethopian barley that protects California's 160-million-dollar annual crop from a dreaded yellow dwarf virus?"

There are less urgent but equally important reasons to preserve biodiversity. Many varieties of food plants not bred by large growers or food companies have superior qualities of taste, nutrition, shape, texture, or heartiness. These are part of our rich biological heritage. Allowing them to become extinct for lack of commercial interest is like losing a traditional art form or a piece of folk music because some company has decided sales aren't high enough. We are all poorer for the loss.

Unfortunately, the trend in the seed business, as in many other businesses, is toward corporate concentration. Large companies have been buying small, regional seed companies. Between 1984 and 1987 in the United States 60 small seed companies were acquired by larger companies and during the same time another 50 small companies went out of business. The result was the same in both cases: heritage varieties of seeds maintained by these small companies were lost. National and international seed corporations want hybrid varieties that yield large volumes and show consistent characteristics. Varieties that don't qualify are deleted from the seed catalogue.

Even more worrisome is the fact that all but one of the world's biggest seed companies is now owned by a petro-chemical company. Ten multinational firms now dominate the world trade in seeds, most of them agro-chemical producers intent on breeding seeds that depend on their pesticides and herbicides. "These firms," says Canadian activist Pat Mooney (Reading 2), "spend much of their resources and time developing seeds that are not pest resistant but pesticide resistant... More and more of the world's farmers are being persuaded to buy these new seeds instead of using their own – that means more and more poisons are dumped on the world's croplands."

With so much money and research being focused on hybrids that can be patented in some countries and therefore monopolized by a single company, heirloom seeds selected by farmers for centuries are being lost quickly. And despite the dangers of genetic uniformity, companies are aggressively pushing their relative few top-performing varieties on hundreds of thousands of farmers across the world.

There is a ray of hope. Non-profit seed-saving groups have been established throughout the world. In Canada, members of the Heritage Seed Program save and share seeds of vegetable varieties threatened with extinction. Dozens of volunteers each cultivate one or more endangered varieties in their garden, then dry and store the seeds. In the United States there is a similar network based in Iowa called The Seed Savers Exchange.

This particular environmental problem – the dwindling of the world's plant genetic diversity – can be reduced to a question of value. When the value of something is measured primarily in terms of commercial value, as it seems to be with seeds, those things without commercial value will be ignored. But when you ignore seeds, they are unable to propagate and eventually become extinct. To prevent this tragic loss, we must adjust our value system. Natural things like plants must be seen as inherently valuable, worth preserving simply because they exist.

Questions and Concerns

READING 1

1. If a new "miracle" seed is developed that increases grain yields dramatically, how can this improve the food supply on one hand while imperiling it on the other?

2. As you read, list the many varieties of plants we commonly rely upon for food which long ago originated in some other part of the world.

3. Why should we care about preserving some obscure bean variety from Mexico or South America?

4. How has terrorism become the latest threat to plant genetic diversity?

The World's Food Supply at Risk

By ROBERT E. RHOADES

THE RANGERS of Ruhunu National Park cannot fathom why we are risking our necks to collect a plant they call "pig's weed." Cursed by local farmers, disdained by cooks, and useful mainly to sorcerers, the weed grows in a part of Sri Lanka inhabited by crocodiles, wild elephants, and terrorists. My job is to watch for crocs and mollify our nervous escorts, whose guns have recently been stolen by rebels.

For all that, my colleague, Balendira Soma Sundaram, the country's chief plant explorer, seems unconcerned. Gripping a pencil in his teeth, he hitches up his rubber boots, adjusts the faded canvas bag on his shoulder, and wades into a lagoon. He scans the shore until he sees the object of our search, a tuft of scraggly weeds half-hidden in the shallows. Reaching out, he plucks a few golden panicles from the stalks, slips the grains into an envelope, and smiles.

"Each time I come here, I find some," he says, happy to have a few

---

ROBERT E. RHOADES, former senior anthropologist at the International Potato Center outside Lima, Peru, wrote "The Incredible Potato" for the May 1982 NATIONAL GEOGRAPHIC. He recently joined the faculty of the University of Georgia in Athens. Photographer LYNN JOHNSON's byline appeared previously in the magazine on "Chicago's Hancock Center," in the February 1989 issue.

more specimens of pig's weed, the wild rice known to scientists as *Oryza nivara,* one of the world's most valuable resources. At least one strain of this endangered species contains an ancient gene that resists grassy stunt virus, a rice pathogen that, sweeping through the paddies of Asia, would be capable of destroying the mainstay of three billion people. Even a 15 percent drop in Asian rice harvests could bring mass starvation.

Balendira and I handle the seeds with due respect, noting the plant's location within the park and labeling a few specimens for shipment to the Philippines. There the rice will be stored in a gene bank in Los Baños, at the International Rice Research Institute. This bunker of concrete and steel is reportedly the strongest building between Tokyo and Frankfurt. Should a new rice virus strike, scientists could obtain *O. nivara* from the bank, attempt to extract a resistant gene, and insert it in other rice varieties to ward off disaster.

Scientists transfer genes between related plants by traditional cross-pollination techniques or, in recent experiments, through genetic engineering. Genetic engineers identify a section of the plant's DNA from which they wish to borrow material. Then, using chemicals, they extract the segment, isolate the gene in a solution, and splice it into the DNA of another plant. In its new home the gene goes to work, repelling insects or fighting diseases just as it had done before.

But biotechnologists cannot *invent* the gene.

That must come from wild sources or from one of the many varieties—which scientists call land races—traditionally bred by farmers. And therein lies the problem. The rice we recovered from Sri Lanka was still there only because it happens to grow in a wildlife reserve. That affords the grass a measure of protection. Outside the park *O. nivara* is disappearing faster than it can be saved. Around the world the same thing is happening to the wild relatives and land races of other major food crops—corn, wheat, and potatoes.

"What people call progress—hydroelectric dams, roads, logging, colonization, modern agriculture—is putting us on a food-security tightrope," said Te-Tzu Chang, head of the rice institute's gene bank, where 86,000 varieties of rice from all over the world are stored. "We are losing wild stands of rice and old domesticated crops everywhere."

Ironically the loss of genetic diversity accelerated with the green revolution of the 1960s. Back then, with the best intentions, scientists developed new "miracle" seeds by carefully crossbreeding plants to increase food production—mostly rice and wheat—in poor nations.

Courtesy of Robert E. Rhoades

The results were dramatic. The new seeds, resistant to insects and diseases, yielded millions of additional tons of grain a year. Indonesia and India, formerly dependent on imports to feed themselves, soon were self-sufficient; India now produces a surplus for export to Sudan and other hungry countries.

The miracle seeds were not perfect, however. Opportunistic insects and viruses mutated and unlocked the genetic resistance of the new seeds. The pests sent scientists scurrying, searching for genes to withstand the threats. They have been successful so far. Meanwhile, the old varieties and wild plants are disappearing from many places, replaced by improved crops that are genetically uniform.

In Sri Lanka, where farmers grew some 2,000 traditional varieties of rice as recently as 1959, only five principal varieties are grown today. In India, which once had 30,000 varieties of rice, more than 75 percent of total production comes from fewer than ten varieties.

The trend toward single-variety monoculture, the planting of one strain instead of many varieties, leaves modern plant breeders little margin for error, says Garrison Wilkes, a professor of biology at the University of Massachusetts and a leading authority on genetic erosion. "The extinction of local land races by the introduction of improved varieties is analogous to removing stones from the foundation to repair the roof," Wilkes adds.

At present rates of extinction, as many as 60,000 plant species—one fourth of the world's total—may be lost or endangered within the next 50 years. Meanwhile, there are more mouths than ever to feed.

WHEN FARMERS began harvesting the first domesticated plants about 8000 B.C., the earth's population was around four million. Today that many people are born every ten days. If the trend continues beyond the year 2000, we will have to grow as much food in the first two decades of the new century as was produced over the past 10,000 years.

The key for meeting that monumental demand for food may be wild plants. Inside each seed is the germ plasm containing the DNA, the genetic code evolved over millions of years in the wild that dictates each plant's development. This "stuff of life" determines a plant's resistance to pests, disease, drought, and similar natural catastrophes. The germ plasm controls the taste, appearance, and preserving qualities of food as well.

"The genes in wild species and old varieties have incalculable value to plant breeders looking for natural resistance to disease and pests," says Gene Saari, a staff scientist with the International Maize and Wheat Improvement Center in El Batán, Mexico, who has introduced improved, high-yield seeds from India to Egypt. "Problem is," he adds, "the old varieties are disappearing as farmers take up modern ones."

Modern farmers prefer the modern varieties, the plants redesigned by genetic scientists who borrow the best attributes from various seeds and blend them into new ones to increase productivity, to meet the taste of consumers, and to provide maximum protein, among other reasons.

But there is a trade-off. By relying on a few crop strains instead of many, farmers open themselves to disaster. In the U.S., for instance, billions of rows of essentially identical corn are planted each year, making the entire crop vulnerable to a single pest or disease.

United States farmers learned that the hard way in 1970, when an unexpected epidemic of corn leaf blight wounded the pride of the world's most agriculturally advanced nation. A virulent new strain of fungus appeared in south Florida that winter and raced north like a killer flu. Since each ear of corn was a copy of every other, there was no margin of safety. The fungus destroyed half the crop from Florida to Texas. Nationwide losses amounted to 15 percent, at a cost of perhaps one billion dollars.

Such disasters are nothing new. Throughout history the sowing of uniform crops has led to a harvest of tragedy.

The collapse of Classic Maya civilization around A.D. 900, some anthropologists speculate, resulted from farmers' planting a mere handful of maize varieties, which were destroyed by a virus. Ireland's infamous potato famine of 1845 started with a fungus accidentally introduced from Mexico. That scourge, spreading through millions of genetically similar spuds, left the Irish without their main food source, and nearly a million people starved to death. A few decades later a fungus wiped out the homogeneous coffee plantations of Ceylon, transforming that island into one of the world's major tea producers. And as recently as 1984 a bacterial disease struck Florida, forcing 135 nurseries to destroy 18 million citrus trees and seedlings.

Luckily, America's bruising by that corn fungus reshaped attitudes toward genetic resources. The National Academy of Sciences set out to assess U.S. vulnerability to crop disaster. The findings were sobering: Half the U.S. wheat acreage is planted in a mere nine varieties, three-fourths of the potato crop in four varieties, half the cotton in three varieties, and more than half the soybeans in six.

Even before the corn blight, agricultural research organizations such as the International Potato Center had been established to broaden and improve the genetic foundation. The potato center, located in La Molina, near Lima, Peru, is one of 13 research groups in the Consultative Group on International Agricultural Research (CGIAR), a consortium conducting some 300 million dollars' worth of research annually. Individual research centers strive to identify, rescue, and preserve wild species and land races of plants. Plant explorers are searching the world over for fresh genetic material.

I ARRIVED in the Soviet Union with a team of scientists sponsored by the U.S. Department of Agriculture (USDA). Our mission was to hunt for seeds in Siberia, but first we stopped in Leningrad to pay homage to one of the giants of modern botanic exploration, Nikolay I. Vavilov.

Vavilov, working in the 1920s and '30s, identified eight specific geographic areas around the world where he believed farmers first domesticated plants. Those areas still have the greatest diversity of major food crops. Directing a staff of 20,000 in more than 400 research stations across the U.S.S.R., Vavilov urged his seed collectors to store plant materials for safekeeping. Long before others realized the value of such collections, Vavilov understood that a cache of diverse genetic material could determine whether a nation's larder was empty or full.

His associates knew it too. During the 900-day siege of Leningrad in World War II, Vavilov's staff faced starvation rather than eat the precious stocks they had painstakingly gathered from the far corners of the earth. In a gesture of stubborn optimism, curators straggled out into the besieged city and planted specimens from their collections. They had to regenerate stocks for the future. During the siege many curators died in the laboratories, their stomach's empty. Surrounding the corpses were the boxes of seeds and sacks of potatoes they had been saving.

Vavilov did not live to witness the suffering of his colleagues. Packed off to prison in Saratov after a scientific dispute with Stalin's pet agronomist, Vavilov died there, accused of spying and agricultural sabotage.

Yet his work continues at the Vavilov Institute of Plant Industry, on an old tsarist estate that is now a favorite location for filming Soviet Sherlock Holmes films.

"Vavilov is not yet as famous as Galileo, but his time will come," Vladimir Krivchenko, then director of the Vavilov Institute, told me. "He admonished us to preserve the plant diversity created over millions of years before it is too late," said Krivchenko, a robust Russian who motions with callused hands that speak of his own involvement with the soil.

A few days later, accompanied by a few of Krivchenko's colleagues, I ventured deep into Siberia, hunting for grass seed with Vasiliy Malofeev, an expert on Siberian plants. Kay Asay, a USDA grass breeder from Logan, Utah, led the American team.

We bounced along in canvas-covered trucks on the only road linking the remote industrial city of Novosibirsk with Mongolia. Each time we spied a promising plant, we stopped. We scoured mountain slopes, storage bins, haystacks—even overgrown graveyards where the dead of the Russian Revolution sleep. All in the search for forage grass.

"Most Americans think this stuff is hayseed," said Kay Asay, showing me a handful of crested wheatgrass. "It's the key to our successful western ranching operations."

Around the campfire one night I learned of our long-standing reliance on Soviet imports. Orchard grass, bromegrass, wild rye, meadow fescue, clover—all came to us from the Soviet Union. Even the pride of the Great Plains, hard red winter wheat, descends from Ukrainian varieties that crossed the ocean with Mennonites in the late 19th century.

Guided through Siberia by Vasiliy and Kay, I stripped handfuls of ripened seed and put them into separate envelopes, carefully labeling each one: "Species: *Bromus* sp. Location: Ust Sema, Katun River, Siberia. Elevation: 1,100 meters. Date: August 13, 1988. Collector: Robert Rhoades." This information would help scientists find the grass again, should it be needed in an emergency.

As the tiny seeds trickled from my palm into the envelope, a thought flashed across my mind: Could one of these grains contain a gene for a minor agricultural revolution? Like the wheat land race from Turkey with a disease resistance worth 50 million dollars annually to the U.S.? Or the Ethiopian barley that protects California's 160-million-dollar annual crop from a dreaded yellow dwarf virus?

Even the most obscure plant can work a minor miracle—on the farm or in the pharmacy. AIDS researchers have found a substance in Chinese cucumber roots that may work against the disease. An extract from the rosy periwinkle of Madagascar has proved effective in treating childhood leukemia. A Mexican yam contributed to the first oral contraceptive. Wild tomatoes, growing in the salty air of the Galápagos Islands, have been used to adapt California varieties to the state's heavily irrigated—and increasingly saline—farmlands. Nature has equipped plants with magical properties that science is only just beginning to discover.

CONSIDER THE MYSTERY of the Mexican bean weevil, an insect with a nasty reputation and cumbersome scientific name. *Zabrotes subfasciatus,* a brown bug about the size of a pencil eraser, destroys as much as 25 percent of the beans stored in Africa and 15 percent in South America. Spraying this major food crop with insecticides might kill the weevil, but it would also harm people. It is safer to repel the bean weevil naturally, by breeding a genetic resistance into its food. That brings scientists like César Cardona into the picture.

For five years Cardona and his colleagues at the International Center for Tropical Agriculture in Cali, Colombia, searched for a bean the weevil would find unpalatable. Ten thousand samples later, Cardona was ready to concede defeat. No cultivated bean, he concluded in a scientific paper, was immune.

"Those little scoundrels tormented me," Cardona recalls, underscoring the memory with a vivid stream of Spanish curses. "I came to hate them!"

A year later, a package of tiny, strange-looking beans appeared on Cardona's desk: They were ugly, blackish brown, trapezoidal, wild beans from Mexico.

"They didn't look like beans," Cardona remembers. "I was laughing at them." But, as a matter of routine, Cardona set them in front of the weevil and waited to see what happened. Nothing happened. The bean's secret armament proved to be a protein, detectable only under chemical analysis, that somehow repulsed the weevil. That protein could be transferred to its cultivated cousins—and was. Now the resistant beans are en route to Africa, a new weapon in the war against famine.

An American plant collector, Howard Scott Gentry, discovered the wild Mexican beans more than 20 years ago, during an expedition to the rugged hills of Guerrero. Exploring on muleback, Gentry spotted vines he had never seen before. He dismounted, gathered a few, labeled them, and forwarded the specimens to the USDA in Beltsville, Maryland. The beans were assigned a plant introduction number—P.I. 325690—and shipped to a plant introduction station in Pullman, Washington. There the beans sat on a shelf, unnoticed and unused, until—in a routine exchange of plant material—the USDA scientists in Pullman shipped a few specimens to their counterparts in Cali.

Gentry, meanwhile, went on with his life, oblivious to the wonders his discovery had wrought. On a visit to his laboratory at the Desert Botanical Garden in Phoenix, I had the immense pleasure of informing Gentry, 87, of the good deed his discovery had set in motion.

A smile spread across Gentry's face. "It makes me very happy that they are headed for Africa," he said, turning back to his plants.

Throughout history people have valued seeds, taking them as prized possessions when they left home. Rice made its way throughout Asia with traveling Buddhist monks. The Pilgrims carried sacks of peas, wheat, barley, and rye, among other seeds, on the *Mayflower*. African slaves often brought a handful of seeds with them to the New World, even if they had nothing else.

Like American music, language, or politics, American agriculture comes from all over. Consider a simple breakfast: Orange juice is squeezed from a fruit that originated in Southeast Asia; toast is made from a grain domesticated in the Near East; hash browns from an Andean tuber, coffee from a wild Ethiopian bush; peach preserves from China.

Trace the lineage far enough, and you learn that the turf at the golf course is Caribbean; the popcorn at the ballpark, Mexican; the Fourth of July watermelon, African; and the amber waves of grain in "America the Beautiful," probably Iraqi.

THE DANGER of importing plants, particularly from a center of diversity, is that they may carry diseases or pests that evolved with them over the centuries. It is for this reason that the borders of the U.S. and many other countries are protected by plant inspection stations. They are the first line of defense against silent invaders that could destroy the standing food supply.

"Are you carrying any plants?" queries an agricultural officer at Miami International Airport.

"No," I reply, handing over my documents. She considers my answer with suspicion, then directs me toward a sign that reads AGRICULTURE. There my luggage rolls through a scanning machine capable of detecting seeds, roots, and stems the way other machines reveal guns. I'm clean.

Deborah Baker, the inspection officer in charge, waves me through. "We deal with all kinds of people. Some try to smuggle endangered orchids. Others bring home a favorite house plant from abroad. Most don't realize how a few seeds in their pockets might introduce a disease."

I ponder how Thomas Jefferson, an incurable seed collector, would react to the fuss. Jefferson knew that America's future depended on new seeds for a viable and varied agriculture, and he encouraged his fellow citizens to import new plants.

"The greatest service which can be rendered any country is to add a useful plant to it's culture," he wrote. In fact, if the United States were forced to live off the plants originating on its own soil, the fare would be slim pickings indeed—sunflower seeds, pecans, strawberries, cranberries, blueberries, and Jerusalem artichokes are among the most palatable items.

As if anticipating Jefferson's advice, Benjamin Franklin, who was serving a stint as Pennsylvania's emissary in England, sent new varieties of seeds to America. By 1827, U.S. consular officers had standing orders to ship home any promising plant they found abroad. In that era, of course, genes were unknown, and there was no concern over introduced diseases or, for that matter, today's baffling new threat to plants and genetic diversity—terrorism.

Maoist guerrillas who call themselves *Sendero Luminoso,* or Shining Path, have overrun a valley high in the Peruvian Andes where my fellow scientists maintain the World Potato Collection, a stockpile of more than 13,000 specimens gathered and cultivated in South America, where the potato originated. Our complacency that the terrorists would not harm their own national heritage was shattered when a busload of workers from the International Potato Center was intercepted by guerrillas in December 1988. One guard was killed. The workers were released, shaken but otherwise unharmed.

A year later three storage buildings at the experiment station were dynamited, forcing the evacuation of the scientific staff to Lima. Suddenly, after years of helping others, the potato center faces an adversary that takes aim upon humanity itself. How do scientists fight that threat?

"I've built this collection through revolutions and earthquakes," says Richard Sawyer, the pistol-packing director general of the potato center. "So I'm not about to kowtow to anybody."

Sawyer, a plain-spoken potato farmer who hails from Maine, has seen worse times. A prisoner of the Germans in World War II, he survived a death march by eating spuds. After that ordeal he became convinced that the plant that saved him could also save the world from hunger. To ensure a healthy potato supply for everyone, Sawyer founded the potato center more than 20 years ago.

Now Sawyer is preparing for what promises to be a protracted struggle. Saving seeds is no longer a matter of simple botanical technique, he tells us. To safeguard the potato collection, we pack samples in bags, stash seeds in airtight metal pouches, and insert plantlets in test tubes—all will be sent to safer regions of Peru and to other gene banks around the world.

"Time is running out," says Carlos Ochoa, a Peruvian who has stalked wild spuds from North America to the tip of Tierra del Fuego. The specter of civil unrest weighs heavily on Ochoa, who sees his life's work imperiled by forces few understand.

ASK ANY AMERICAN schoolchild where the country's most valuable national treasure is stored, and the answer will surely be Fort Knox. But the greatest wealth may be tucked away at a USDA facility on the campus of Colorado State University at Fort Collins, Colorado. Here, in an unassuming two-story building, 228,000 samples of seeds containing trillions of genes are cached in the National Seed Storage Laboratory, the central reserve bank of germ plasm in the U.S. The storage vaults are protected by ultramodern security systems—microwave scanners and infrared sensing devices. Recently Congress authorized twelve million dollars to expand this national gene bank.

"No one knows when a pest will mutate and attack our crops," says Steve Eberhart, director of the laboratory. "Most Americans don't know how close we came to being a food importing nation during the 1970 southern corn blight. Now if we hear that a new corn blight has attacked China, our scientists can prepare to fight it by the time it gets here."

Dr. Eberhart leads me through rooms where his colleagues are testing seeds for viability and preparing them for storage. Some 3,000 seeds of each plant are kept on hand for eventual distribution to any plant breeder who needs them.

Some of these seeds can be kept viable for decades, preserved in stainless steel cryovats in the frigid vapor of liquid nitrogen kept at minus 196°C. From Washington, D.C., to Hawaii, scientists in 19 germ-plasm research stations have kept alive seeds and cuttings from thousands of plants, some of which arrived in the New World more than a hundred years ago.

"The world is different today from what it was in the age of the great plant hunters," says Dr. Eberhart. "Before, we could go to a plant's center of origin and find what we needed. Today many varieties in our collections are extinct in their natural habitats. And many governments have closed their borders to collecting."

WHO OWNS the genetic resources of a plant? Is germ plasm a natural resource—like oil or timber—to be exploited, controlled, or sold? Or is it the common property of all humankind? Like automobiles or toasters, some genetically engineered crops can be patented. This means that genetic material taken from land races and wild species could make large profits for seed companies.

"The risk with plant patent laws," explains Canadian Pat Roy Mooney, a farmers-rights activist with the Rural Advancement Foundation International, "is that the seed companies obtain monopoly profits, whereas the farmers and countries that donated the genes receive nothing.

"For example, among tens of thousands of structural genes in a corn variety nurtured over the centuries by farmers, only a handful generally are altered by commercial breeders to produce a new hybrid. Does that give the breeder a right to patent it and reap the profits—profits sometimes earned by selling the hybrid back to the country where the genes originated?"

Mooney argues that commercial seed companies should be required to contribute a fraction of their profits to an international fund that would subsidize traditional farmers.

Back in the U.S., I got another view from Donald Duvick, then senior vice president for research at Pioneer Hi-Bred International, Inc., a company with annual sales of 870 million dollars.

"Over 90 percent of Pioneer's seed is sold in the U.S., Canada, and Europe," says Duvick. "Only a small fraction goes to poor farmers in developing countries." According to Duvick, it has taken a huge investment—more than ten years of time and tens of millions of dollars—to develop some of the successful new varieties.

"It's a bit like crossing a house cat with a wildcat," says Duvick. "You don't automatically get a big docile pussycat. What you get is a lot of wilderness that you probably don't want lying on your sofa."

Attempts to control the flow of plant wealth are nothing new. During the spice wars of the 17th and 18th centuries the Dutch uprooted groves of nutmeg and clove trees—to keep prices high and to cut their competitors out of the market. And a sticky disagreement persists over rubber trees the British transplanted from Brazil in 1876, transforming millions of acres in their Asian colonies into lucrative rubber plantations.

DESPITE the verbal bombshells over seed sovereignty, I found an astonishing degree of cooperation among scientists and governments. In 1988 the USDA distributed 30,000 samples of seeds and cuttings to some 80 countries. If war or famine destroys native crops, international seed banks can replace them. Valuable potato varieties were lost to Bolivia after workers at the Belén seed bank rose up in protest over low wages and ate the national collection. The International Potato Center sent duplicates of the most important varieties to replenish the supply.

Many of Cambodia's indigenous food plants were lost in the Khmer Rouge reign of terror in the late 1970s. When that strife finally subsided, the International Rice Research Institute dipped into its reserves, returning more than 400 rice varieties to Cambodia, so the country could make a new start.

The spirit of reciprocity has been captured in a traditional Asian saying: "You cannot pick up a grain of rice with one finger alone."

MANY HANDS ARE AT WORK in the U.S., where local seed hunters search old fields, rocky hillsides, and abandoned farms to identify and save long-forgotten seeds. Often laymen, these enthusiastic collectors exchange information and germ plasm, hoping their efforts will provide a needed dose of variety to America's kitchen gardens and small-scale farms.

The largest North American seed-gathering network is Seed Savers Exchange, founded and run by Diane and Kent Whealy from their 140-acre farm in the Amish country near Decorah, Iowa. The inspiration came in 1975, when Diane's grandfather, Baptist John Ott, gave them seeds from Bavaria that her family had brought to this country four generations before.

The Whealys—realizing that such "heirloom" seeds were being lost all over the United States—took up the challenge of developing a grass roots organization to save and swap such seeds. In the past 15 years the network has grown from a handful of people to some 5,000 backyard gardeners who maintain more than 12,000 heirloom fruit and vegetable varieties.

Today the heirloom gardening movement is spilling over to living history farms, where modern crops are being replaced by those of the appropriate era. These historical varieties are becoming as much a part of the farms as heirloom reapers or grinding mills.

At Thomas Jefferson's Monticello outside Charlottesville, Virginia, I find John Fitzpatrick, director of the Center for Historic Plants. "We have over 500,000 visitors a year," he told me. "And we're finding that more and more they prefer to take home as a souvenir a plant Jefferson himself once tended instead of a dust-catching knickknack."

STORING SEEDS as heirlooms is better than letting them vanish, but many plants are best preserved in their original habitats.

"Most countries in Asia have centers for the conservation of art, music, and religion, but only a few have them for seeds," says Gerry Jayawardene, the proud head of Sri Lanka's new Plant Genetic Resources Centre. A few other places offer similar encouragement. In Mexico's Sierra de Manantlán Biosphere Reserve I found a small patch of *teosinte,* the closest wild relative of maize, growing in an area defended by park rangers fending off illegal herders.

But in Texas I discovered miles of pavement and screeching jets where valuable stands of wild grapes once flourished. Now the land is covered by the Dallas-Fort Worth International Airport. Perhaps the loss of a few grapes is a small price to pay for progress, until one thinks back to the 19th century, when an American louse, *Phylloxera,* brought the European wine industry to its knees.

Accidentally introduced into Europe in the 1860s, the louse ravaged thousands of vineyards before a solution was found. Since the American louse attacked European roots, someone finally hit upon the idea of grafting the European vines to American rootstock. Perhaps the New World roots had evolved a genetic resistance to the pest? The plan worked. The American roots kept the louse at bay, and the grateful Europeans were soon drinking wine again. But if a new breed of phylloxera should appear, to whom would the world turn today?

Rock stars and activists decry the loss of rain forests, and an admiring public leaps to the defense of pandas and snow leopards. But who speaks for a weedlike potato too bitter to eat or scraggly rice that spoils the pot?

These thoughts ran through my mind the last time I visited Luther Burbank's Gold Ridge Farm in Sebastopol, California, and saw the three remaining acres of the plant wizard's spread in a state of disrepair. Overgrown with weeds and brush, sandwiched between a housing project called Burbank Orchards and a cemetery, the old farm had been threatened by development. It seemed a fitting symbol of our society's indifference to genetic diversity. I was happy to learn later that the development had been abandoned after community organizations complained, and the farm was being revived.

From 1885 until his death in 1926, Burbank conducted his pioneering experiments with plants here. I saw his arbor of seedless grapes, his hardy Chinese-hybrid orange trees. Burbank bred hundreds of new plants, among them improved varieties of squash, plums, tomatoes, lilies, poppies, and roses. Yet he never obtained a patent. Not until four years after his death did Congress pass a plant patent act, which protects certain new varieties.

"You can almost see the old man

working out there, pruning that old Royal walnut he planted in 1885," says Bob Hornback, a local horticultural historian who, working with the Western Sonoma County Historical Society, is among those helping to restore Gold Ridge to its former glory.

"THE SCIENTISTS BLAME the loggers. Loggers blame the settlers. Settlers blame the government. All of them are full of contradictions. Too many people who want too much."

The speaker, Achmad Jahja Kostermans, talks rapidly as he pads barefooted through the botanical garden in Bogor, Indonesia. Though he is 84, I have trouble keeping pace with Professor Kostermans, the greatest living botanist of the Asian tropical rain forest. The disappearance of food crops, he says, is part of a larger problem. The genetic diversity of the entire plant community is eroding, and nowhere is the rate of destruction greater than in parts of Southeast Asia.

"The chain saw sounds like an angry beast eating up the forest—*rrrrgh…rrrghhhh!!!*" Kostermans cries, mimicking the sound.

"Protected areas!" Kostermans laughs. "They exist only on paper in poor countries. Once I found a new tree species in a protected area in western Java. Only one of its kind, and loggers cut it down in a week."

This aged botanist appreciates the practical value of tropical forests. They saved his life when he was a prisoner of the Japanese in World War II. Kostermans, conscripted to help build the bridge over the River Kwai, survived on forest plants, using them for food and medicine. Of the 20,000 other prisoners, 18,000 died. When the war ended, a grateful Kostermans dedicated his life to the study and preservation of natural flora. He has also adopted students, so that others can continue his work.

He points to one of his protégés, a young Frenchman named Jean Marie Bompard, a researcher for the International Board for Plant Genetic Resources. "I am the past," says Kostermans, with an uncharacteristic softness in his voice. "He is the future."

Bompard's future will be busy. It is said that Kostermans has gathered enough plant material to keep another 50 scientists working for 50 years, just cataloging and testing it. Somewhere in Kostermans' collection may be the wonder seed of tomorrow or the raw material for a miracle drug that will cure AIDS—no so farfetched when you consider that about 25 percent of U.S. prescription drugs come from plants.

More than even life-saving drugs or food, plants and seeds carry a powerful symbolism in many cultures. From the Hopi of Arizona, for example, I learn that seeds represent a sacred link to their past, handed down through ceremonies from generation to generation. To them, each seed represents hope for the future as well.

MY JOURNEY ENDED in southeastern Turkey, somewhere near the Garden of Eden. In this mystical land, bounded by the Tigris and Euphrates Rivers, Adam was doomed to work "accursed" soils that yielded the "brambles and thistles" of Genesis.

Here too, according to Sumerian epics, the legendary Utnapishtim landed his ark after a monstrous flood, finding a home for "the seed of all living things." This was one of several far-flung areas where prehistoric humans made their first bold experiments with wild plants, sowing seeds to grow their own food.

Even though the forests have vanished and the surrounding hills are denuded, the grain bursts forth with the spring, as in all the seasons before. I found scattered stands of wild wheat and barley.

But the cycle may end soon, when the valley is swallowed by another flood. If all goes as planned, Turkey will build 21 dams on the Tigris and Euphrates by the early 21st century. Perhaps that is a fitting symmetry. The land between the rivers will be green again, an irrigated area the size of the Netherlands and Belgium combined. But the Garden of Eden will be lost forever.

READING 2

1. The author of this article writes, "a community's accumulated farming wisdom gives it strength and self-reliance." What does she mean?

2. What are Pat Mooney's criticisms of large international agricultural research organizations like Consultative Group on International Agriculture Research (CGIAR) and the International Board for Plant Genetic Research (IBPGR)?

3. If the Uruguay Round of the General Agreement on Tariffs and Trade (GATT) is accepted by signatory countries, and patent law is applied to plant life, what will the implications be for Third World farmers?

Profile

Seeds of Change: Pat Roy Mooney

Canada's own prophet of "a people-based agricultural revolution that will allow the poor to eat and the earth to survive."

by Penny Sanger

In 1985, in the parliament of Sweden, Pat Roy Mooney—Prairie born-and-bred father of five and a Winnipeg high school dropout—was awarded the Right Livelihood Award, the Alternate Nobel Peace Prize. His "vision and work forming an essential contribution to making life more whole, healing our planet and uplifting humanity" were cited in the ceremony.

Mooney had realized in the early 1970s that Western-style "development" was fast eroding the world's great storehouse of food crop seeds. His work since then has been to preserve what's left of the vast, disappearing reservoir of strengths, resistances and physical characteristics that constitutes the world's plant genetic resources.

For Mooney this requires much more than preserving the different varieties of apples and potatoes our grandparents grew. Farmers down the ages have "selected out" different strains of plants, to resist pests or disease or weather. In tropical agriculture, where almost all our basic food crops originate, growing varied types of a crop like sorghum or maize or rice could mean the difference between a harvest and starvation. So a community's accumulated farming wisdom gives it strength and self-reliance. Pat Mooney saw that by replacing these folk varieties with Western-bred high-yield plants the world was losing plant types that could never be bred again, farmers were being driven to rely on pesticides and herbicides instead of their own seeds, and farming communities lost their independence.

Mooney's obsession with seeds and genetic diversity has been a 20-year one-man struggle. He has faced the combined hostility of scientists and agriculture and aid establishments, who decried his lack of formal training. The Canadian International Development Agency (CIDA) and most Canadian policymakers pooh-poohed his revelations for years. Some still claim that advances in genetics and biotechnology outpace the loss of "landrace," or historic, farm-bred varieties of seeds. Others allege his campaign was rooted in opposition to the

Penny Sanger is a member of the Forum's *editorial board.*

Green Revolution of the 1960s—which is still feeding millions. His reply is usually an avalanche of statistics that add up to the loss of Third World farm small-holdings, and increased dependency on Western technology.

"Pat's crazy," his closest friends say. But they come close to revering the power of his argument, the truth of his facts, and the combination of tenacity and humour with which he works "the system."

"We have already allowed 75 per cent of genetic diversity in the 20 key food crops to be lost forever, mostly during the last 50 years," Mooney warns.

His independence is legendary. Nevertheless he and his organization, Rural Advancement Foundation International (RAFI), are committed, enthusiastic workers with other non-governmental organizations.

Along [the] trail Mooney [has] stirred up the wrath of agrochemical giants like Ciba-Geigy and ICI, leaders of world trade in fertilizers and chemicals, as well as the seeds they breed to depend on those chemicals.

He uncovered the North's malignant business and political hold over the very people that international agricultural institutions are supposed to be helping—the world's poor. He took on the agricultural development system and won. Now he wants to turn it right side up so that it can do the job intended for it. Most particularly he and his colleagues at RAFI want to slow down the North's headlong rush into international patenting of plant varieties and other life-forms.

"There can be no true land-reform—no true agrarian justice of any kind—and certainly no national self-reliance, if our seeds are subject to exclusive monopoly patents and our plants are bred as part of a high-input chemicals package in genetically uniform and vulnerable crops," Mooney has said.

A Teenage Activist

To sit down across a desk from Pat Mooney is like plunging into a river in full spate. A torrent of charts, tables, maps and documents pours out. It's not until he stops to check something, and his head drops toward the document to read that you remember he's almost blind.

As a child he was also afflicted with polio. He and his brother Terry used to play marathon games of Geography ("Atlanta-Azerbaijan-Norway-Yangtze," etc.) during the 12-hour train journeys back and forth from their home in Saskatoon to a doctor's office in Winnipeg.

But it was maps that imprinted the shape of the world on his young mind, and stretched his vision beyond the Prairie cities where he and his family lived.

"The walls of our house were covered with them," he recalls. "*National Geographic* maps, in the hallways, the kitchen, the bedrooms, the living room." A knowing grandmother who gave them the subscription helped start the boys on parallel but very different international careers. Terry is in the policy division of the Canadian International Development Agency.

The Geography games gave way to Kelvin High School in Winnipeg. Terry became president of the history club, "the best, most interesting club." Pat, the younger brother, joined the United Nations club, got elected its president, and eventually amalgamated the two. As he grew he polished this skill of patiently working within a system, shaping it toward his goals.

His lifelong commitment to the idea of the United Nations began at this time. Somewhere he read that 10 cents was all it took to cure a child of yaws. He convinced his high-school principal to hold a special school assembly on October 24, UN Day, and started campaigning for 10-cent contributions. He raised $4,000. Next day he presented an astonished Winnipeg UNICEF office with the money.

> *"We have already allowed 75 per cent of genetic diversity in the 20 key food crops to be lost forever, mostly during the last 50 years."*

It was then the sixties, and the hyperactive teenager, already losing his sight, was bored with high school. He dropped out, won a trip to a youth assembly in Vienna, and on the way home stopped off in Rome to visit a friend who worked in the UN's Food and Agriculture Organization (FAO). He met the director-general.

"I was a long-haired high-school dropout...in those days that opened all doors," he laughs. He returned home as a FAO youth ambassador to the world. Today it is FAO—a renewed and restructured FAO—that should, he believes, lead badly needed world reform in agricultural development.

Destroying Diversity

A chance discovery in Kenya that Canadian wheat is made up of germplasm, or genes, from all over the world turned Pat into a seed addict. Countries such as Kenya itself, now so impoverished it relies on imported grains to feed its people, are in fact the progenitors of wheat strains that once helped make the Canadian Prairies rich.

"I went straight to the Winnipeg library when I got home and did the research...It was true." This simple fact about wheat, followed by the years of lobbying and research to come, convinced him that food, the basis of life, has become a high-stakes trading game.

Another young activist was already heading toward this conclusion from his base in Pittsboro, North Carolina. Cary Fowler, sociologist and co-author (with Frances Moore Lappe and Joe Collins) of the seminal food-as-politics book *Food First*, was a seed researcher at Rural Advancement Foundation. The young men met at a conference and soon became close allies in the struggle to put food—and the plants that produce it—on the international development agenda.

Their 1990 book *Shattering: Food, Politics and the Loss of Genetic Diversity* exposes the controlled partnership between seeds and chemicals. It shows that in the agriculture business 10 companies dominate world trade in seeds and most of them are also leading manufacturers and traders of pesticides, herbicides and other chemicals.

Mooney explains, "These firms spend much of their resources and time developing seeds that are not pest resistant, but *pesticide* resistant. They are very deliberately developing and marketing seeds that are dependent on their own products—pesticides and herbicides. More and more of the world's farmers are being persuaded to buy these new seeds instead of using their own—that means that more and more poisons are dumped on the world's croplands." RAFI has identified at least 65 research organizations that focus on breeding herbicide/pesticide tolerance into food crops.

This is a long way from the farmer-proved method which, over millennia, selected and bred untold numbers of food crop varieties with specific characteristics to withstand the vagaries of weather, weeds and disease. Mooney is convinced that the potential of primary producers to feed the world's poor can only be met if their traditional skills and accomplishments are used, and if the international agricultural development establishment is re-built from the ground up to include them.

"Until poor farmers in the South have a voice in these bodies they can't start taking control of their lives, making their own decisions...," Mooney says.

An example is Ethiopia, whose farmers are inheritors of one of the world's greatest stores of plant diversity. Famine, poverty and development assistance have decimated this richness. Today the brilliant Ethiopian plant geneticist Dr. Malaku Worede leads an international consortium of NGOs, including Canada's Unitarian Service Committee, to help them recover what's left.

Many farmers in Ethiopia (and in other parts of the poor world) were convinced by Western aidworkers to abandon their traditional varieties of crops for the new high-yield but less reliable strains. The fearful 1984-85 famine brought increased shipments of new "high-response" varieties of wheat, barley and sorghum. The result is that today there are only four main varieties of wheat growing on the Ethiopian plateau, where once hundreds of indigenous strains, each known for its special attributes, flourished. Nearly half the country's sorghum harvest comes from imported seed.

"It's like trading the Elgin Marbles for the Rolling Stones," says Mooney.

Farming communities have developed and maintained different types of food seeds over the millennia in response to changing conditions. It's an insurance against crop disease and climate variation. If a year of drought or grasshoppers came, some varieties would succumb, others would endure. Each variant was known and treasured for its particular strength and characteristics.

Plant breeders all over the world have always used these "folk" varieties as a store of genetic material from which to fashion new designer seeds. Rarely, if ever, do they pay their originators for the privilege. Now, belatedly, they see the world's stock of gene resources declining fast. At the same time, global warming and the short life-spans of many new varieties—dependent as they are on ever-changing formulations of fertilizers and herbicides—means a constant search for new survival characteristics. The retrieval and re-fashioning of these tested Third World folk genes is a race against extinction. The small farmer, accustomed at harvest time to select for re-planting the fattest heads of grain and those that had stood up strongly to pests and climate change, has either switched to the new commercial seeds or gone to the city. In the Punjab, where the "improved" seeds were introduced with great fanfare, nearly one-quarter of small-holdings failed between 1970 and 1980. With them were lost their community's seed varieties, the products of literally thousands of years of field-testing and human farming experience.

Corporate Manoeuvres

The corporate powers that flogged the new varieties did so behind a screen of international bodies that were very deliberately outside the United Nations system, Mooney says. In that Cold War period, "...private US foundations and the US government systematically undermined the development agenda of many UN organs including the FAO...." They created, with Ford and Rockefeller foundation money, a string of international agricultural research centres (IARCs) around the world.

These were charged with developing bigger and better varieties of major food crops—thus CIMMYT in Mexico works on maize and wheat, IRRI in the Philippines on rice, CIAT in Columbia works on tropical agriculture, ICRISAT in Hyderabad on semi-arid tropical agriculture, and ICARDA in Syria on agriculture in dry areas.

Somehow the skills and experience of the local and regional farmers they work among got pretty well forgotten in all this Western-run research and experimentation.

"And the IARCs are continuing to draw peasant farmers into the mould required by the Western food system in the developed world," Mooney emphasizes. Following the Canadian wheat example, they bring the South's biological treasures to industrialized countries, for use in northern labs and northern farmers' fields.

Together with relatively unknown but powerful bodies like the Consultative Group on International Agricultural Research (CGIAR) and the International Board for Plant Genetic Resources (IBPGR) these research centres had a collective annual budget far in excess of the UN's Food and Agriculture Organization. Mooney comes close to accusing them of allowing the agro-chemical corporations to run wild, ignoring the richness and talents of Third World agriculture in order to make a buck.

Banking and Biology

Pat Mooney knows his palmy days won't last much longer. He's watching the industrialized world's latest moves to cream even more profit out of Third World agriculture. The Uruguay Round of the General Agreement on Tariffs and Trade (GATT) proposes that worldwide patent and intellectual property legislation be applied to plant life. This would ensure victory and huge additional profits to northern breeders and agro-chemical enterprises. Third World farmers would have to pay for the right to use genetic plant material originally grown on their own farms—the products of their own skills and management.

It could lead to a Third World "intellectual property" debt of $60 billion annually, as much as the Third World's current yearly debt repayments.

"If the GATT proposal goes through," says Mooney, "we'll see companies like ICI licensing the plant varieties for which it has patent 'protection' to, say, a company in Kenya. That company will sell them to local farmers and the royalties will flow back to ICI. The Kenyan company will do what it can to gain patent control over local folk strains. In both cases the farmer will pay more."

The U.S. and Japan argue that the Third World's failure to adopt the same intellectual property protection as is available in the U.S. amounts to an unfair trading practice. They want Third World member countries to pay for the plant varieties the West has already taken from them, and developed or stored in northern gene banks. They estimate the annual loss of such "royalties" to northern enterprises runs between $40 and $70 *billion.*

In his 1990 Massey lectures, *Biology as Ideology,* R.C. Lewontin talked about how high-yielding hybrid corn was first developed in the 1920s in the southern U.S. He showed that careful selection could have produced equally strong, varied and prolific strains. The motive for developing hybrids was entirely commercial.

"What are said to be fundamental discoveries about the nature of life often mask simple commercial relations that provide a powerful impetus for the direction and subject of research," he began. "Self-reproduction presents a serious problem to someone who wants to make money developing new varieties of organisms...."

"[Agro-chemical] firms spend much of their resources and time developing seeds that are not pest resistant, but pesticide resistant. They are very deliberately developing and marketing seeds that are dependent on their own products..."

"...We have known the truth of the matter for the last 30 years. The fundamental experiments have been done and no plant breeder disagrees with them....By the method of selection plant breeders could, in fact produce varieties of corn that yield quite as much as modern hybrids. The problem is no commercial plant breeder will undertake such investigation and development because there is no money in it."

It is selection, the basic tool of all farmers through the ages, that produced the rich folk seed diversity that breeders now are trying to recover.

Powerful players in agricultural research, such as Canada's International Development Research Centre, are also worried about the trend to globally enforced intellectual property rights. The authors of "A Patent on Life" (IDRC 1991) warn that once commercial interests dominate, most agricultural research resources will be spent on the crops that bring in the most money—those grown on large acreages in favoured environments. "Minor crops that may be of vital importance to resource-poor farmers in diverse and often harsh environments will inevitably be neglected by commercial plant breeders, leading to ever greater inequities.

Questions and Concerns

READING 3

1. Where do you think the Mostoller Wild Goose bean might have come from before it was found by the Mostoller family? Why is this bean such a miracle today?

2. What defines an "endangered" seed?

Apple's Seeds

Back to the future in the Heritage Seed Program garden

BY SUSAN HALDANE

When a marauding groundhog moved into Heather Apple's vegetable garden last year, there was more at stake than a good autumn feed and the winter's freezer stock. Blossoming with such rare and oddly named plants as 'Djena Lees Golden Girl' tomatoes, 'Moon and Stars' watermelons and 'Cherokee Trail of Tears' beans, Apple's garden was not just a collection of fresh salad greens – it was, in fact, living history.

Apple – "When I was a kid I hated my name, but now I love it" – is president of Canada's Heritage Seed Program (HSP). Her rolling acreage near Uxbridge, Ontario, is also home to the programme and to more than 100 seed varieties she has personally adopted. Her foster family of seeds awaits the next sowing in clear, airtight glass jars arranged in rows on a cupboard shelf, and inside every one of those jars is a story. Some of the stories are mere novellas about a unique genetic makeup. Others, however, are epics telling of rich family histories and struggles to survive and adapt.

One of the most famous dramas in Apple's library is titled 'Mostoller Wild Goose' bean. Long ago, the story runs, a man named Joseph Mostoller decided to increase his income by building a timber mill at the junction of Wells Creek and Stoneycreek, near his home in Pennsylvania. Joseph's two sons John and David were fighting in the Civil War, so Mostoller began his labour alone. By the time the mill was ready to start operation in 1865, the two boys had returned. That year, an early winter forced a number of ducks and geese to pause in their migration for a weather delay on Stoneycreek. One of the geese had the misfortune to stray into the millrace within view of the two brothers, who shot it and turned it over to their mother, Sarah.

As she cleaned the bird, Sarah noticed that its crop was full, and a quick autopsy revealed some unusual white beans splashed with pink and reddish brown markings. Mother Mostoller set the beans on a windowsill to dry and planted them the following spring. The gift of the wild goose turned out to be ideal for baking and soup making. It was passed on to neighbours and handed down through generations of the Mostoller family. Today, 81-year-old Ralph Mostoller is the custodian of the

Courtesy of Susan Haldane

family heirloom, and he personally answered Apple's request for some seeds to add to her collection.

The story – and the beans – would probably have remained a family secret if not for people like Heather Apple. Her Heritage Seed Program is dedicated to preserving varieties of vegetables threatened with extinction and to saving and sharing both the stories of the seeds and the seeds themselves. Members volunteer to adopt an endangered variety, grow it in their garden, save the seeds and list them in a catalogue published each winter. For a membership fee of $10 plus the cost of postage, any member can receive a starter kit of seeds. Launched in 1984 under the flag of the Canadian Organic Growers (COG), the HSP now has 270 members, and its catalogue for the 1990 growing season will include almost 180 varieties of vegetables offered by members who grew and preserved seed last year.

The tale of the Mostoller bean is told and retold in seed-exchange publications as one of the great success stories. But the fear of such an heirloom dying along with the last of the green-thumbed Mostoller family members is just one of the threats the Heritage Seed Program battles in its defence of genetic diversity.

"When I first heard about this," says Apple, "I was really concerned because I've always loved diversity. I have the disease of loving seed catalogues. Each year, my order gets bigger and bigger, and I'd need a troop of gardeners to plant everything that I'd like to get."

Publications of the HSP and of its American counterpart, the Seed Savers Exchange, identify four enemies of genetic diversity. Family heirlooms like the 'Mostoller Wild Goose' bean can be lost in a single generation, but even more heritage varieties disappear whenever a small, regional seed company goes out of business or is taken over by a larger company. Statistics published by the Seed Savers Exchange show that between 1984 and 1987, 60 small seed companies were taken over and another 50 folded. As a result, 943 varieties available in 1984 could no longer be purchased five years later. The national seed corporations and commercial producers prefer hybrids that promise consistency and high yields, so it becomes harder and harder for backyard gardeners to buy the time-honoured, open-pollinated (nonhybrid) varieties. Finally, Apple sees a further menace in plant breeders' rights legislation, which would allow patenting of new varieties. Critics of the legislation argue that corporate complaints of unfair competition could halt government-sponsored plant-breeding projects and leave the market in the hands of the multinationals. Others fear that corporate breeders might gain control of an older variety by making minor genetic alterations and patenting it.

These four threats are "narrowing the genetic base of all our food plants," says former HSP president Alex Caron of King City, Ontario, "and that makes us more and more vulnerable to disaster. That's what happened in the Irish potato famine." In the 1840s, all the potatoes grown in Ireland had a similar genetic base. When a blight struck, it wiped out crops of this staple throughout the country. Hundreds of thousands of people starved, and many more immigrated to the New World. Caron says this tragedy has had more recent echoes – a near disaster for hybrid corn crops in the 1970s, for example – proving that people have not yet learned the moral of the story: genetic diversity is important. The corn was almost all the same hybrid and thus similarly vulnerable to disease.

Heirlooms, on the other hand, represent a mixture of traits. What one cultivar is susceptible to, another may resist. There may come a day when plant breeders will need to produce a plant resistant to a new disease or a change in the environment. "Who knows what qualities these varieties may have that have never been called upon?" Apple asks. So she and Caron and others like them have become self-appointed keepers of these mysterious bundles of genetic potential. The task of the seed savers can be compared to saving tins of food that have lost their labels: you don't know what's in them, but they may come in handy someday.

Caron's own interest in heritage vegetables was piqued in 1979, when an acquaintance asked to borrow enough garden space to grow an obscure potato. He called it 'Austrian Crescent' because it came from Austria and had the long, narrow shape of a crescent moon. Come harvesttime, the acquaintance didn't show up, and Caron was left holding the potato sack. The 'Austrian Crescent' became the first of what is today a collection of more than 200 unusual and heirloom potatoes.

There are no heirloom plants in Apple's family. In fact, plants were a low priority in her life until her parents moved to their isolated 60 acres in Scugog Township, north of Whitby, Ontario. Growing up in Toronto, Apple had had her own small patch in the family's backyard garden, a space she devoted to flowers – Apple remembers a batch of daffodils that struggled to a mature height of about three inches after she buried the bulbs too deep. But that hardly mattered; she was far more interested in the fauna her garden offered than in the flora. She smuggled jars of worms, caterpillars and sow bugs into the house, where she lovingly cared for them as long as she could keep them a secret. When it came time for university, Apple chose to study zoology.

"At that time in my life, I thought that plants were the inferior half of the nature kingdom," Apple says. "I had friends who were in botany, and I thought, 'Well, isn't it sad, they've certainly chosen something a little duller than animals.'"

But "dull" took on new meaning after she spent several months counting fruit flies for a university geneticist. The next summer, when her parents asked her to take on their vegetable garden as a full-time job for the season, she considered it the lesser of two evils.

"It turned out that I really, really loved it. Looking back, I wonder why, because our garden was just laced with twitch grass, and we didn't understand about twitch grass in those days. I knew nothing about gardening. But the sheer miracle of growing vegetables was very rewarding. You know, planting a potato and having a potato plant and digging it up to find potatoes underneath, or planting a carrot seed and having it grow into a real carrot, just like the carrots you buy in the supermarket, only better.

"I recently looked through some slides I'd taken of that first garden. I grew Hubbard squashes that year, and we had an abundant harvest – about a hundred squashes. I must have taken about 20 slides of those squashes – close up, from a distance, from every angle, with each member of the family, even with each dog sitting beside a Hubbard squash. I was so amazed, so wonder-struck, that we had produced so many squashes."

Miracle or no miracle, Apple still wasn't convinced that the story of her life was to have gardening as its central theme. When COG held a conference in 1984 to discuss threats to genetic diversity, Apple was concerned and interested but not committed. "At that point, I was determined to have nothing more to do with gardening, except as a nice hobby. I was going to be an artist, so I didn't get too involved."

The Heritage Seed Program ran for a couple of years under Caron's guidance, but he found his volunteer position as president, secretary, treasurer and sole heritage gardener too much to handle. By the time he went looking for new recruits in 1987, Apple had undergone another career conversion. She'd finally accepted gardening and raising food for her parents as a full-time job and was ready to open her home and garden to the programme.

Now beginning her third year as president – and indeed, the only executive member – of the HSP, Apple admits she had no idea when she started how much work would be involved or how the seeds would come to dominate her life. Surrounding her as she speaks, racks of drying seeds and the materials for preserving them fill every flat surface of a room that does multiple duty as library, workshop, studio and office. Books on gardening and seed saving crowd the zoology texts on her shelves. Apple's seed museum is in the corner – a large, white cupboard about the height of a hoe handle and the width of two rake heads. On the floor, boxes full of dry beans labelled 'Horsehead' and 'Jacob's Cattle' await shelling.

"I've put all the boxes of beans kind of politely over to one side, because it's sometimes too depressing if I look at everything all at once," she says. "But a lot have disappeared. Every time my parents sit down to watch TV, I give them bags of beans to shell. And when I go out to meetings, I take my beans with me. Actually, I usually take the peas to meetings, because shelling beans makes too much noise."

Added to the work of growing and saving her own seeds is the huge job of organizing and overseeing the national network of growers. Apple, who admits she had "never even turned on a computer" before she started, now publishes three magazine-style newsletters a year as well as the December catalogue of seeds available for members. She also teaches classes on seed saving and acts as spokesperson for HSP on radio and television and in print. Her future plans won't make the volunteer workload any more managable: she is considering compiling a catalogue of all the fruit, nut and berry varieties available in Canada, and she has made it a goal to search out vegetables unique to this country, such as the 'Mac Pink' tomato, an early-ripening type developed at McGill University's Macdonald College.

"I'm called upon to do a whole range of things, and there are certain things that I'll plunge in blindly and do, and I feel fairly happy about doing them. But there are other things that I'm kind of horror-struck about doing. I'm recognizing that I'm going to have to start hiring people and writing for grants, and that's something I don't feel very confident about." (An initial grant – $50,000 over five years from the Weston Foundation – "fell out of the sky," Apple says, when a representative for the food-empire family called to ask if the programme wanted any money.)

But when Apple begins to feel overwhelmed by the responsibilities of pioneering a grass-roots seed exchange, she finds inspiration in Seed Savers Exchange, the American organization founded by Kent Whealy in the early 1970s. Apple modelled the HSP after Whealy's seed-swap system and hopes to follow in his footsteps by establishing a permanent home for the seed collection.

If Whealy is the HSP's spiritual leader, his book *The Garden Seed Inventory* (Seed Saver Publications, Decorah, Iowa, 1985) is its bible. Chapter and verse, the good book of heritage gardening lists virtually every unusual nonhybrid seed sold by almost every seed company in the United States and Canada. If a seed is available from only one or two sources, it is considered endangered. For example, both 'Moon and Stars' watermelon and 'Cherokee Trail of Tears' bean list two sources, while 'Mac Pink' tomato can be bought only from W.H. Perron in Laval, Quebec. But listed with the 'Cherokee Trail of Tears' under "Bean/Dry/Snap" are 13 cultivars no longer available commercially. Unless someone somewhere has been growing them and saving the seed, no one will ever again taste 'Blue Lake Number Seven,' 'Champagne,' 'Miss Kelly' or 'Sultan's Emerald Moon.' Of the 5,291 nonhybrid varieties Whealy lists, 67 percent are available from only one or two companies. For Heather Apple, that means they're on the endangered list, and she struggles to save as many as she can.

A tour through her garden is like a stroll through a century-old seed catalogue. There against the north fence are the 'Cherokee Trail of Tears' beans, believed to have been raised by the Cherokee people and taken with them when they were uprooted by white settlers in the 1800s. Sprawling in the centre of the one-acre space is a vine feeding a dozen 'Moon and Stars' watermelons, their green-black skins speckled with the bright yellow markings that give them their name. 'Moon and Stars' was thought to be extinct until a television interview with Kent Whealy turned up a gardener who had been keeping them in rural Missouri. Off in a corner are the 'Djena Lees Golden Girl' tomato plants, heavy with fruit.

The first heritage seed Apple grew was 'Jacob's Cattle' bean, a type of dry bush bean with aliases including 'Trout,' 'Coach Dog' and 'Dalmation.' She still keeps them, and she holds out a handful – white with a block of maroon and a network of maroon flecks. The bean is said to have been discovered in native American ruins in the southwestern United States and supposedly dates back to about 1200 A.D. A neighbour gave Apple the beans along with a recipe for baking them. Since she knew she wouldn't find them in her Stokes or Ontario Seed Company catalogues, Apple vowed to save her own seeds to plant the next year. Her collection was under way. Last summer's garden included 14 varieties of heritage tomatoes, six types of squash and pumpkins and a dozen kinds of beans.

But preserving seed for the history books is not as simple as setting some seeds on a windowsill the way Sarah Mostoller did. There are at least as many pitfalls as there are days in the growing season, and they come at all stages in the process. Of course, a heritage garden can fall victim to any of the evils that stalk a conventional garden – like last year's groundhog. It's a frustrating enough invasion for any gardener, but Apple had just planted 12 of the 16 'Mostoller Wild Goose' beans Ralph Mostoller sent her, so she knew there was an epic history growing out there. From seeding time until harvest, "I'd practically go out and pray over them every day," she says. "Fortunately, the groundhog was so taken up with my squash, he never moved over to the beans."

Apart from the usual challenges of disease and climate, a heritage garden presents some unique problems. To keep the strains pure, Apple is forced to hand-pollinate varieties that are not self-pollinating. With her squash, for example, she would have to race the bees to the blossoms at the crack of dawn if she did not tape the flowers closed each evening. The next morning, she removes the tape, rubs the pollen from the male blossom onto the female and tapes the female up again, precluding future bee visits. When the flower dries, she ties string or ribbon around the stem so that she will know at harvesttime which vegetables contain pure seed. "Last year was a real drag, because we had a plague of cucumber beetles. They chew their way into the blossom overnight, so I had to protect every female flower with little bags I made."

Once the seeds are harvested, they must be thoroughly dried to prevent mould during storage. Apple dries hers on racks at room temperature before putting them into airtight jars and shelving them in the cupboard. Then, to keep the seeds fresh and viable, they must be "grown out" every three to five years. If she wants to store the seeds any longer, she dries them with silica gel and stocks them in a freezer.

The jars and cupboard are another lesson learned the hard way. The first year Apple saved seeds, she stored the precious products in paper envelopes on top of a cabinet. She was sure there was no way a mouse could get at them but didn't bargain on a chipmunk breaking in and making off with the better part of her collection.

Such disasters aside, Apple says she's convinced she made the right choice in adopting the Heritage Seed Program and its tiny artifacts as her own. And she is optimistic that the seeds have a future as well as a history. "There's a tidal wave toward genetic uniformity, but fortunately, there's a wave – seed exchanges and growing public interest – moving in the opposite direction. Also, there's a new breed of seed companies that really values genetic diversity." The more gardeners support these companies by buying unusual seeds, she says, the more they ensure that a broad and varied genetic base will be preserved.

One final and very compelling reason for saving heritage varieties is their taste. Often, the new varieties that are the brainchildren of large seed companies are bred more for their ability to offer consistent yields and survive long-distance transportation than for flavour, usually the top priority for backyard gardeners. Alex Caron swears by 'Mrs. Morrell's' yellow-fleshed potatoes. Ralph Mostoller wouldn't cash in his beans for anyone's, and even Apple occasionally get to eat one of her museum pieces.

Among the heirlooms and simple oddities she has grown, some favourite tomato varieties are 'Prudens Purple' (a gift from Caron), 'Djena Lees Golden Girl' and 'Camp Joy' cherry. The taste of these unusual tomatoes, she says, "is different from 'Sweet 100,' a modern hybrid, but I love 'Sweet 100' too. We're a real sugar generation. We've all been brought up on sugar, and we all like sweet things, so a lot of people judge the taste of tomatoes, especially cherries, by the taste of 'Sweet 100.' But I also enjoy the old-fashioned, really rich tomato taste."

Apple remains an unusual gardener, however, because eating the fruit of her labour is not a top priority. "My job is to grow food for my parents, right? So I grow all these varieties of heirloom tomatoes, and they're not allowed to go out in the garden and pick a tomato. You see, I need the seeds out of them first. So I go out and pick the tomatoes, and I very nicely cut them in half and scoop out the seeds, and then I give my parents the remainder. Once I have enough seeds, they can have whole tomatoes, but they have to do their tithing."

Offering up vegetable sacrifices to the HSP didn't prevent the production of a huge batch of chili sauce in the Apple household this fall, and that's what Heather wants. Part of the job of the programme, she believes, should be to teach people how to use these unusual vegetables and make them part of their own family stories.

"If people have a rich and diverse culture, then their lives will be enriched. Similarly, their lives will be enriched if they have a rich and diverse horticulture. That's something the Heritage Seed Program is trying to promote: as well as just preserving the varieties, it's helping people to enjoy them, and that's why our motto is 'preserving and enjoying our horticultural heritage' – because when people enjoy it, then it's worth saving."

Ancient Foods for Future Diets

Have You Had Your Spelt Today?

BY RACHEL O'NEAL

Though we often take the food we eat for granted, the abundance we Americans enjoy today is largely due to technological advances of only a few years ago. During the 1960s, agricultural technologists developed high-yielding rice and wheat crops that could be grown just about anywhere. But with revolutionary zeal, these crops were often transferred to inappropriate areas – rice in the California desert, for example, further strains an already stressed water supply.

These "advances" also encouraged mono-cropping (the planting of large areas with single plant species), causing the eclipse of many native crops by modern hybrids. Limiting ourselves to a few major food staples, like wheat, corn, soy and rice, has left crops more vulnerable to damage from insects and weather. A more diverse agricultural mix ensures that, in the event one crop is devastated, others survive.

Health experts are beginning to suspect that this lack of diversity in our food supply may also be contributing to health problems. Allergies or sensitivities to wheat, for example (the most common), may be due simply to its overabundance in our diets.

Though we tend to look for technological quick-fixes in the face of modern dilemmas, one answer to such problems may lie in the re-emergence of ancient foods, some of which have already started a comeback: spelt, kamut, teff, quinoa and amaranth.

The Grandaddy: Spelt

Mentioned in the Bible, spelt, a type of wheat, was grown in Mesopotamia more than 9,000 years ago. Amish farmers brought spelt with them from Europe when they immigrated here 100 years ago, and have grown a variety suitable only for cattle ever since. There are only two known food-grade varieties suitable for human consumption.

Wilhelm Kosnopfl, founder of Purity Foods of Okemos, Michigan, brought the first food-grade spelt to America from Europe just five years ago and placed it with certified seed growing farmers in Michigan to maintain a pure seed stock. Other farmers in Minnesota, Wisconsin, Illinois, Indiana and Ohio also now grow spelt.

Higher in protein and iron than common wheat, spelt is well tolerated by most wheat sensitive individuals. It is an excellent wheat substitute in bread-making, producing loaves virtually indistinguishable from common wheat bread. It is also being used for cereals, pasta, pizza crusts, bread sticks and even tortillas.

> "Limiting ourselves to just a few major food staples has left our food crop more vulnerable to damage from insects or climatic changes."

Mysterious Kamut

Kamut, another ancient wheat, has a curious origin. In 1949, Earl Dedman, a U.S. airman stationed in Portugal, received some kernels of a giant "wheat," supposedly from Egypt. He sent the seeds back to his father, a Montana farmer who grew it briefly as a novelty. Calling the grain, "King Tut's wheat," he exhibited it at the State Fair and distributed seed in jars, but failed to stimulate interest in it as a crop.

Years later, Bob Quinn, another Montana farmer, remembered seeing the unusual wheat at the State Fair as a child. "Because of its size, it was probably peddled in Egypt as a giant wheat of the pharaohs," suggests Quinn. He and his father looked all over Montana for the ancient wheat and managed to find one jar of seed. They started cultivation in 1978, calling their crop "kamut," an ancient Egyptian name for wheat. In fact, says Quinn, kamut evolved in the fertile crescent area around 4000 B.C., and still exists as a non-commercial crop in the northern Mediterranean. The Quinns currently contract with about 30 growers in North Dakota, Montana, Saskatchewan and Alberta. About 4,000 acres of kamut were planted in 1992.

Kamut is a relatively low-yielding grain, susceptible to disease and particular about climate. Such limitations, says Quinn, "will prevent it from sweeping the country, rather keeping it regional. Which is good, because it does well in a region where we can't grow corn or soybeans. It'll never replace the main grains, but it increases diversity."

Kamut is high in protein, magnesium, potassium and zinc. Like spelt, kamut is well tolerated by the majority of wheat sensitive individuals. It makes a tasty pizza crust and a breakfast cereal that some contend tastes better than oatmeal.

> "Allergies or sensitivities to wheat, corn, soy or rice may be due to their overabundance in our diets."

The "Mother Grain": Quinoa

Central to the ancient Inca civilization of South America, quinoa (keen-wah, or "mother grain") was suppressed by the conquering Spaniards in the 16th century. Cultivated since 3000 B.C., this staple became a minor crop, grown only in remote areas.

Quinoa's spinach-like leaves are important greens in the South American diet. Quinoa's high protein content is so balanced in amino acids that the National Academy of Sciences has called it "the best source of protein in the vegetable kingdom." Quinoa's ability to thrive in high altitudes with low rainfall, subfreezing temperatures and poor soil make it ideal for marginal cropland.

Steve Gorad and Don McKinley started Quinoa Corporation in America and first attempted cultivation of Quinoa in the U.S. in 1982 in the Colorado Rockies, spurred by a love of the ancient food that Gorad originally sampled in South America. It is difficult to grow in North America, however; currently, most quinoa commercially sold is imported from South America, where it grows naturally at 14,000 feet in the Andes mountains.

Quinoa Corporation buys from about 10,000 native farmers directly. "I have a problem paying middlemen while poor farmers don't get full value for their crop," says Quinoa Corporation president David Schnorr. "The farmer's entire crop may be only a few hundred pounds. Most grow 80 percent of it for their own use, and sell the rest."

Quinoa Corporation currently offers only one of 2,200 possible varieties. "We chose the 'real' – meaning "royal" in Spanish – variety because it was the most prevalent," explains Schnorr.

Quinoa is quick and easy to prepare, cooking in just 15 minutes. It is excellent served in a cold salad with chopped vegetables, and is also good in casseroles, stews and stuffings. It makes a great pudding and can be used in pastries, cakes and quick breads.

Immortal Amaranth

When Cortez conquered the ancient Aztecs in Mexico, he banished their staple, amaranth, as a commercial crop. Cultivation in the wilds of South and Central America and Mexico prevented amaranth from vanishing off the face of the Earth.

Wild amaranth (meaning "unfading" in Greek because the flower lasts so long) was discovered growing in remote Mexico by a botanical research team in 1972. Seeds were brought to this country and experimentally grown at various universities. Today, besides being grown in more than 20 countries, amaranth is grown in the U.S. in Colorado, Kansas, Minnesota, Montana, Nebraska and Wyoming. It is also grown in Africa in semi-arid regions due to its drought tolerance. It can even be raised in a home garden for its tasty greens.

"When you buy amaranth, you get more nutrients for the money spent than in most other foods," claims Larry Walters of the American Amaranth Institute. "But nutrition, whether you have great taste or not, is a hard sell to consumers, who eat foods mostly for pleasure."

Amaranth is high in iron and protein and, like quinoa, supplies all the essential amino acids, making its protein complete. Its sweet, nutty flavor produces tasty cookies and quick breads. Substituting about one quarter amaranth when cooking breakfast corn meal greatly enhances its nutrition and adds a delightful crunch to every bite.

Ethiopian Teff

Indigenous to the highlands of Ethiopia, teff has been grown for 4,000 years by

COMPARATIVE NUTRITION INFORMATION

Grains are a good source of protein, fiber, carbohydrates and B vitamins, and contain an average of 200 calories per 2-ounce serving. Amaranth and Teff are high in calcium, supplying approximately ten percent of the U.S. Recommended Daily Allowance (RDA) per 2-ounce serving. Here's how they stack up in other areas:

RECOMMENDED DAILY ALLOWANCE	PROTEIN	CARBO-HYDRATES	FAT	FIBER	IRON	THIAMIN	RIBO-FLAVIN	NIACIN
	45g	n/a	n/a	25g	18mg	1.5mg	1.7mg	20mg
Grams per 2-ounce serving					*% RDA per 2-ounce serving*			
AMARANTH	8g	38g	4g	9g	24%	3%	7%	4%
KAMUT	10g	39g	1g	*	13%	17%	4%	16%
QUINOA	7g	39g	3g	3g	29%	7%	13%	8%
SPELT	8g	42g	2g	5g	13%	24%	7%	24%
TEFF	7g	41g	1g	8g	25%	15%	4%	4%
BROWN RICE	1g	44g	2g	2g	5%	15%	3%	14%
OATS	10g	38g	4g	6g	15%	29%	4%	3%
WHOLE WHEAT	7g	40g	1g	7g	10%	14%	4%	15%

*Information not available

Source: U.S. Department of Agriculture. Chart: John Kanzler. Information compiled by Alyssa Burger.

Ethiopian farmers who use it to make injera, a bread which is a staple in their diet. Though teff was never suppressed by conquistadors, it wasn't because no one tried. Traditionally, conquests were carried out at harvest time. Burning the fields before harvest and destroying the seed stock effectively starved out the soon-to-be-conquered culture. However, farmers could carry enough of teff's small seeds in their pocket for survival. "With a small handful, you can grow a whole field," explains Wayne Carlson, the first teff grower in this country. "Teff gave a resiliency to the Ethiopian culture that prevented domination."

Carlson got interested in teff in the 1970s while doing research in Ethiopia. Back in the states, Carlson missed Ethiopian food and began to consider growing teff domestically. He set up his business, Maskal Teff, in Caldwell, Idaho, and in 1985 brought the first domestically grown teff crop to market.

The sunny days, cool nights and reliable irrigation of Idaho and Eastern Oregon copy teff's native climate well enough for successful cultivation. Maskal Teff grows several types of teff out of the hundreds of varieties possible. Currently, brown or white teff are available commercially as grain or flour.

A nutritional powerhouse, teff is high in iron and calcium. It is well tolerated by people who can't eat traditional grains. Delicious as a hot breakfast cereal, teff can also be used in soups, stews, pilafs and quick breads.

Helpful Resources

These ancient foods, often organically grown, can be purchased at most natural foods stores, in the health food sections of some supermarkets, and via mail order through the following companies:

- *Allergy Resources Catalog*, P.O. Box 888, Palmer Lake, CO 80133/(719) 488-3630.
- *Goldmine Natural Food Co.,* 1947 30th St., San Diego, CA 92101/(800) 475-FOOD.
- *Mountain Ark Trading Company,* 120 South East Ave., Fayetteville, AR 72701/(800) 643-8909.
- *Wysong Corporation,* 1880 N. Eastman Rd., Midland, MI 48640/(517) 631-2314.

For more information, contact:
- *American Amaranth Institute,* 384 Seely St., Bricelyn MN 56014/(507) 653-4377.
- *Kamut Association of North America,* 2161 Meyers Avenue, Escondido, CA, 92029/(619) 747-3008.
- *Maskal Teff,* 1318 Willow, Caldwell, ID 83805/(208) 454-3330.
- *Purity Foods, Inc.,* 2871 West Jolly Road, Okemos, MI 48864/(517) 351-9231.
- *Quinoa Corporation,* P.O. Box 1039, Torrance, CA 90505/(310) 530-8666.

RACHEL O'NEAL *is a freelance writer living in Portland, Oregon.*

Discussion for Biodiversity

1. Should companies be allowed under "intellectual property" laws to own the genetic resources of a plant? Agro-companies argue that the millions spent on plant genetic research justifies patent monopolies over the hybrid varieties they develop; that it's the only way to earn a fair return on their investment. Those opposed to patenting genetically "engineered" plants say that often these large companies only make small alterations to the genetic makeup of plant varieties raised for centuries by farmers, yet the farmers are not compensated. Indeed, the farmers are more likely to be sold the new genetically altered seed – and the pesticides and herbicides needed to maintain it – at a premium price by the seed companies. One suggestion is that the seed companies should contribute a fraction of their profits to support traditional farmers across the world who are preserving heirloom species. What do you think?

2. Farm activist Pat Mooney says, "There can be no true land-reform – no true agrarian justice of any kind – and certainly no self-reliance, if our seeds are subject to exclusive monopoly patents and our plants are bred as part of a high-input chemicals package in genetically uniform and vulnerable crops." How do land reform, agrarian justice, and self-reliance relate to monopoly patents on seeds? Discuss the connections.

3. Should we feel uncomfortable because petro-chemical companies own all but one of the world's 10 largest seed companies? Why have these corporations bought up seed companies?

4. Discuss how your class might tap into the heirloom seed network and "adopt" several endangered vegetable varieties.

5. Compare the characteristics of the food we eat that is derived from wheat (bread, pasta, cereals) to the characteristics of the various ancient grains mentioned in Reading 4. Why would people eat these uncommon grains? What are the benefits?

Mining Update

Update on Raven/Berg/Johnson *Environment*, p.187

Updated information on some environmental aspects of coal mining, as discussed on page 187 of *Environment*, was not able to be placed in the first printing. The authors would like to direct the reader's attention to the following changes:

column 1: The 90,000 mining fatalities recorded in the 20th century should be viewed in a historical perspective. There have been great improvements in reduced fatalities due to the efforts of mining companies and to federal regulators in enforcing safety standards. According to the U.S. Department of Labor, 54 coal miners died during 1992, which is the second lowest total in history. Annual accidental deaths of coal miners numbered from 2000 to 3000 in the earlier part of the century.

column 1: There are more than 2000 deaths per year in the United States resulting from black lung disease (a form of pneumoconiosis), although it is difficult to determine precise numbers. Improved ventilation equipment and other safety measures may result in a lower incidence rate of this disease in the future; the matter is under study by coal companies and government agencies.

column 2: As discussed on page 330 of *Environment*, the 1977 Surface Mining Control and Reclamation Act (SMCRA), requires coal companies to restore areas that have been surface-mined, in order to prevent the past practice of large open pits left after surface mining. SMCRA also requires that topsoil be replaced after mining, and that procedures to control erosion and comply with water quality criteria are established.

Note: Look for more coverage of reclamation of surface-mined lands in the second edition of *Environment*.

Please complete the other side of this form, detach and return.

FOLD HERE AND TAPE CLOSED

_____
_____
_____

PLEASE PLACE POSTAGE HERE

Please circle the manual you use:
Northeast
Southeast
Mid Atlantic
Great Lakes
Midwest
Southwest
Northwest
Canada

Return to:
Regional Environmental Issues Manuals
Julie Levin Alexander, Senior Editor
Saunders College Publishing
The Public Ledger Building
620 Chestnut Street, Suite 560
Philadelphia, PA 19106

Professor and Student Comments
Your Feedback Counts!

The *REGIONAL ENVIRONMENTAL ISSUES MANUALS* are a "living" project which will be growing and changing with editions, like our environment. We at Saunders College Publishing are delighted to be pioneering such new and unique supplements for our environment textbooks, as a way of bringing regional issues closer to your home. Your feedback counts! Please detach, complete and return this form so we can continue to update our manuals and cover the environmental issues that are important in your region.

Your Name _____ Are you a Student _____ or Professor _____ ?

School Affiliation and State _____

Address _____

City _____ State _____ Zip _____

Phone Number   Home: _____   Office: _____

Which Regional Environmental Issues Manual do you use? (Circle one)

Northeast	Southeast	Mid Atlantic	Great Lakes
Midwest	Southwest	Northwest	Canada

What issues in this manual were most interesting and useful to you in class? _____
_____

What issues can be improved? _____
_____

Are there issues/topics which are not included which should be included in the next edition? _____
_____
_____

Are there issues/topics which should be deleted from the manual? _____
_____

What additional sources (articles, readings) would you recommend for inclusion in the next edition? _____
_____
_____

Do you feel your state is placed in an appropriate region, one in which other states of the region have similar environmental problems and concerns?

Yes _____   No _____   If not, what region should your state be in? _____

How would you rate the Regional Environmental Issues Manual as a free supplement for your text:

____ Excellent and very informative   ____ Needs to be improved in the future
____ A good regional supplement to our text ____ Poor
____ Average   ____ Other (Please explain) _____

For Instructors:
Would you be interested in working on a revision of this or other Regional Environmental Issues Manuals? Yes _____ No _____

Other Comments: _____
_____
_____

Please complete the other side of this form, detach and return.

FOLD HERE AND TAPE CLOSED

PLEASE PLACE POSTAGE HERE

_____
_____
_____

Please circle the manual you use:
Northeast
Southeast
Mid Atlantic
Great Lakes
Midwest
Southwest
Northwest
Canada

Return to:
Regional Environmental Issues Manuals
Julie Levin Alexander, Senior Editor
Saunders College Publishing
The Public Ledger Building
620 Chestnut Street, Suite 560
Philadelphia, PA 19106

Professor and Student Comments
Your Feedback Counts!

The *REGIONAL ENVIRONMENTAL ISSUES MANUALS* are a "living" project which will be growing and changing with editions, like our environment. We at Saunders College Publishing are delighted to be pioneering such new and unique supplements for our environment textbooks, as a way of bringing regional issues closer to your home. Your feedback counts! Please detach, complete and return this form so we can continue to update our manuals and cover the environmental issues that are important in your region.

Your Name _____ Are you a Student _____ or Professor _____ ?

School Affiliation and State _____

Address _____

City _____ State _____ Zip _____

Phone Number Home: _____ Office: _____

Which Regional Environmental Issues Manual do you use? (Circle one)

      Northeast   Southeast   Mid Atlantic   Great Lakes

      Midwest   Southwest   Northwest   Canada

What issues in this manual were most interesting and useful to you in class? _____
_____

What issues can be improved? _____
_____

Are there issues/topics which are not included which should be included in the next edition? _____
_____
_____

Are there issues/topics which should be deleted from the manual? _____
_____

What additional sources (articles, readings) would you recommend for inclusion in the next edition? _____
_____
_____

Do you feel your state is placed in an appropriate region, one in which other states of the region have similar environmental problems and concerns?

Yes _____ No _____ If not, what region should your state be in? _____

How would you rate the Regional Environmental Issues Manual as a free supplement for your text:

____ Excellent and very informative   ____ Needs to be improved in the future

____ A good regional supplement to our text   ____ Poor

____ Average   ____ Other (Please explain) _____

For Instructors:
Would you be interested in working on a revision of this or other Regional Environmental Issues Manuals? Yes _____ No _____

Other Comments: _____
_____
_____